100 Challenging Calculus Problems

Complete Solutions

Jeff Wang

Published by HiTeX Press

For permissions and other inquiries, write to:
P.O. Box 3132, Framingham, MA 01701, USA

Contents

Introduction

Calculus, as a field, stands as one of the primary pillars of modern mathematics. It provides the analytical tools needed to navigate the complexities of continuous change and variation. Its impact spans multiple disciplines, serving as the mathematical foundation for physical sciences, engineering, economics, and beyond. In pursuit of deeper understanding and advanced competence, students are often encouraged to engage not merely with calculus's theoretical underpinnings but also with its practical applications through problem-solving.

This book, "100 Challenging Calculus Problems: Complete Solutions," offers a comprehensive exploration of calculus's problem-solving landscape. It aims to extend the knowledge and sharpen the analytical skills of readers who seek to challenge themselves beyond the standard curriculum. Each problem has been meticulously chosen to reflect a diverse range of topics within the realm of calculus. From limits and continuity to advanced techniques of integration and series expansions, the problems presented herein are intended to provoke thoughtful consideration and foster significant learning progression.

The intent of this compilation is neither to provide mere drilling exercises nor to reiterate already familiar themes. Instead, this text aspires to engage students in critical thinking, compelling them to navigate through problems that require ingenuity and comprehensive understanding. In addressing these challenges, students are expected to develop a mastery over calculus that is both robust and versatile.

The solutions provided in this book are intended to guide and educate by elucidating the reasoning and methodologies that underpin each answer. A thorough approach has been adopted to ensure clarity and completeness, maximizing the pedagogical value of each solution. Readers are encouraged to actively engage with the problems and compare their approaches with the detailed solutions presented.

Throughout this work, we maintain a focus on precision and elegance, reflecting the inherent nature of calculus itself. Students and educators alike are invited to utilize this text as a resource for deep academic inquiry, intellectual challenge, and the pursuit of excellence in advanced calculus.

It is the hope of this book that through engagement with these problems, learners will gain not only an enhanced understanding of calculus but also an appreciation for the discipline's beauty, rigor, and vast potential for application. Thus, each chapter serves as a step toward the reaffirmation of calculus's integral role in both academic and professional spheres, aiding in the cultivation of a mathematically rigorous mindset.

As you embark on this exploration of advanced calculus problems, we wish you success in your endeavor to elevate your understanding and extend the boundaries of your mathematical knowledge.

Chapter 1

Limits and Continuity

This chapter examines the fundamental concepts of limits and continuity, vital to the understanding of calculus. It presents a series of challenging problems designed to test and deepen the understanding of these essential topics. The problems focus on exploring the behavior of functions as they approach specific points or infinity, and evaluating whether these functions are continuous. Readers will develop critical analytical skills and gain insights into the intricacies of limits and continuity through detailed and complete solutions.

1.1   Evaluate $\lim_{x \to 0} \frac{\sin(x)}{x}$.

The limit $\lim_{x \to 0} \frac{\sin(x)}{x}$ is a fundamental expression in calculus, particularly in the study of trigonometric limits. This limit is essential in the derivation of derivatives for trigonometric functions and plays a pivotal role in various applications across physics and engineering. Understanding this limit requires a careful analysis of the behavior of the sine function as the angle x approaches zero. The problem involves determining the exact value of the ratio of $\sin(x)$ to x as x becomes infinitesimally small, approaching zero from both the positive and negative directions.

To evaluate the limit $\lim_{x \to 0} \frac{\sin(x)}{x}$, we employ several mathematical techniques, including geometric interpretations, the squeeze theorem, and series expansions. Each method provides a different perspective on why the limit equals one, reinforcing the robustness of this fundamental result.

$$\lim_{x \to 0} \frac{\sin(x)}{x}$$

Solution:

We aim to determine the value of $\lim_{x \to 0} \frac{\sin(x)}{x}$. To do so, we will explore multiple approaches to solidify our understanding and ensure the correctness of the result.

1. Geometric Interpretation and the Squeeze Theorem:

Consider a unit circle with radius 1. Let x be an angle measured in radians, and $0 < x < \frac{\pi}{2}$. We can associate $\sin(x)$ with the height of the corresponding point on the unit circle, and x with the length of the arc subtended by the angle x.

In the unit circle:

- The length of the arc corresponding to angle x is x.

- The area of the sector formed by angle x is $\frac{1}{2}x$.

- The area of the triangle formed by the radius and the sine of the angle is $\frac{1}{2}\sin(x)$.

- The area of the triangle encompassing $\tan(x)$ is $\frac{1}{2}\tan(x)$.

From geometric considerations, we obtain the following inequalities:

$$\cos(x) \leq \frac{\sin(x)}{x} \leq 1$$

As x approaches 0, $\cos(x)$ approaches 1. Therefore, by the Squeeze Theorem, since $\frac{\sin(x)}{x}$ is "squeezed" between $\cos(x)$ and 1, both of which approach 1, it follows that:

$$\lim_{x \to 0} \frac{\sin(x)}{x} = 1$$

2. Series Expansion (Maclaurin Series):

The Maclaurin series expansion for $\sin(x)$ is given by:

$$\sin(x) = x - \frac{x^3}{3!} + \frac{x^5}{5!} - \cdots$$

12

Dividing both sides by x:

$$\frac{\sin(x)}{x} = 1 - \frac{x^2}{3!} + \frac{x^4}{5!} - \cdots$$

As x approaches 0, the higher-order terms $(x^2, x^4, \ldots)$ become negligible. Therefore:

$$\lim_{x \to 0} \frac{\sin(x)}{x} = 1$$

3. L'Hospital's Rule:

L'Hospital's Rule applies to limits of the form $\frac{0}{0}$ or $\frac{\infty}{\infty}$, where direct substitution is indeterminate. Since $\sin(0) = 0$ and x approaches 0, the limit is initially of the form $\frac{0}{0}$.

Applying L'Hospital's Rule, we differentiate the numerator and the denominator:

$$f(x) = \sin(x) \quad \Rightarrow \quad f'(x) = \cos(x)$$

$$g(x) = x \quad \Rightarrow \quad g'(x) = 1$$

Thus, the limit becomes:

$$\lim_{x \to 0} \frac{\sin(x)}{x} = \lim_{x \to 0} \frac{\cos(x)}{1} = \cos(0) = 1$$

4. Using Trigonometric Identities:

We can also use the identity $\sin(x) \approx x$ for small angles x measured in radians. This approximation is valid because, as x approaches zero, the difference between $\sin(x)$ and x diminishes rapidly.

Therefore:

$$\lim_{x \to 0} \frac{\sin(x)}{x} \approx \lim_{x \to 0} \frac{x}{x} = 1$$

Conclusion:

Through geometric reasoning, series expansion, L'Hospital's Rule, and trigonometric identities, we consistently find that:

13

$$\lim_{x \to 0} \frac{\sin(x)}{x} = 1$$

This result underscores the close relationship between the sine function and linear behavior near the origin, a cornerstone concept in calculus that facilitates the differentiation of trigonometric functions and the analysis of periodic phenomena.

1.2 Determine the Continuity of the Function $f(x)$ at $x = 0$.

Determine the continuity of the function $f(x)$ at $x = 0$, where $f(x)$ is defined as follows:

$$f(x) = \begin{cases} x^2 \sin\left(\dfrac{1}{x}\right), & \text{if } x \neq 0, \\ 0, & \text{if } x = 0. \end{cases}$$

Solution

To determine the continuity of $f(x)$ at $x = 0$, we need to verify the following three conditions:

1. **Existence of $f(0)$:** Verify that $f(0)$ is defined.

2. **Existence of the limit:** Compute $\lim_{x \to 0} f(x)$.

3. **Equality of limit and function value:** Check whether $\lim_{x \to 0} f(x) = f(0)$.

Step 1: Verify that $f(0)$ is defined.

From the definition of $f(x)$:

$$f(0) = 0.$$

Therefore, $f(0)$ exists.

Step 2: Compute the limit of $f(x)$ as $x \to 0$.

We need to evaluate:

14

$$\lim_{x \to 0} x^2 \sin\left(\frac{1}{x}\right).$$

Observation:

- The function $\sin\left(\frac{1}{x}\right)$ oscillates between -1 and 1 infinitely often as $x \to 0$.

- The amplitude of the oscillations is multiplied by x^2, which approaches 0 as $x \to 0$.

Applying the Squeeze (Sandwich) Theorem:

We know that for all $x \neq 0$:

$$-1 \leq \sin\left(\frac{1}{x}\right) \leq 1.$$

Multiplying all parts of the inequality by x^2 (which is always non-negative):

$$-x^2 \leq x^2 \sin\left(\frac{1}{x}\right) \leq x^2.$$

Thus, we have:

$$-x^2 \leq f(x) \leq x^2.$$

As $x \to 0$, both $-x^2$ and x^2 approach 0. Therefore:

$$\lim_{x \to 0} -x^2 = 0, \quad \lim_{x \to 0} x^2 = 0.$$

By the Squeeze Theorem:

$$\lim_{x \to 0} f(x) = 0.$$

Step 3: Compare the limit to $f(0)$.

From Step 1:

$$f(0) = 0.$$

From Step 2:

$$\lim_{x \to 0} f(x) = 0.$$

Since:

$$\lim_{x \to 0} f(x) = f(0),$$

all three conditions for continuity at $x = 0$ are satisfied.

Conclusion:

The function $f(x)$ is *continuous* at $x = 0$.

Additional Note:

To illustrate why the function is continuous despite the oscillations of $\sin\left(\dfrac{1}{x}\right)$, consider plotting $f(x)$ near $x = 0$. The oscillations become increasingly compressed and their amplitude diminishes due to the x^2 factor, effectively smoothing out the function at $x = 0$.

Answer:

The function $f(x)$ is continuous at $x = 0$.

1.3   Find $\lim_{x \to \infty} \left(1 + \frac{1}{x}\right)^x$.

In this problem, we seek to determine the limit of the function $\left(1 + \frac{1}{x}\right)^x$ as x approaches infinity. This expression is fundamental in calculus and is intrinsically connected to the number e, which is the base of the natural logarithm. Understanding this limit not only provides insights into exponential growth processes but also lays the groundwork for more advanced topics such as continuous compound interest and differential equations. The challenge lies in evaluating the behavior of the function as x becomes arbitrarily large and discerning the value to which the function converges.

To evaluate $\lim_{x \to \infty} \left(1 + \frac{1}{x}\right)^x$, we proceed by taking the natural logarithm of the function, which simplifies the exponentiation, allowing us to apply limit laws and properties of logarithms effectively. Let L denote the limit:

$$L = \lim_{x \to \infty} \left(1 + \frac{1}{x}\right)^x$$

Taking the natural logarithm of both sides gives:

$$\ln L = \ln\left(\lim_{x \to \infty}\left(1 + \frac{1}{x}\right)^x\right)$$

By the continuity of the natural logarithm function, we can interchange the limit and the logarithm:

$$\ln L = \lim_{x \to \infty} \ln\left(\left(1 + \frac{1}{x}\right)^x\right)$$

Using logarithmic identities, specifically $\ln(a^b) = b \ln a$, we rewrite the expression:

$$\ln L = \lim_{x \to \infty} x \cdot \ln\left(1 + \frac{1}{x}\right)$$

To evaluate this limit, we consider the behavior of $\ln\left(1 + \frac{1}{x}\right)$ as x approaches infinity. For large x, $\frac{1}{x}$ becomes very small, and thus $\ln\left(1 + \frac{1}{x}\right)$ can be approximated using the Taylor series expansion of $\ln(1 + h)$ around $h = 0$:

$$\ln(1 + h) = h - \frac{h^2}{2} + \frac{h^3}{3} - \cdots$$

Substituting $h = \frac{1}{x}$, we have:

$$\ln\left(1 + \frac{1}{x}\right) \approx \frac{1}{x} - \frac{1}{2x^2} + \frac{1}{3x^3} - \cdots$$

Multiplying this approximation by x:

$$x \cdot \ln\left(1 + \frac{1}{x}\right) \approx x\left(\frac{1}{x} - \frac{1}{2x^2} + \frac{1}{3x^3} - \cdots\right) = 1 - \frac{1}{2x} + \frac{1}{3x^2} - \cdots$$

As x approaches infinity, the terms $\frac{1}{2x}$, $\frac{1}{3x^2}$, etc., tend to zero. Therefore, the limit simplifies to:

$$\lim_{x \to \infty}\left(1 - \frac{1}{2x} + \frac{1}{3x^2} - \cdots\right) = 1$$

Thus, we have:

$$\ln L = 1$$

To solve for L, we exponentiate both sides:

$$L = e^{\ln L} = e^1 = e$$

Therefore, the limit is equal to e, the base of the natural logarithm.

$$\lim_{x \to \infty} \left(1 + \frac{1}{x}\right)^x = e$$

This result is pivotal in calculus, as it defines the number e through a fundamental limit involving exponential growth. The derivation employs properties of logarithms and limits, showcasing the interplay between algebraic manipulation and calculus concepts to arrive at a precise and significant mathematical constant.

1.4 Compute $\lim_{x \to 0} \frac{e^x - 1}{x}$.

The problem requires evaluating the limit of the function $\frac{e^x - 1}{x}$ as x approaches 0. This expression involves the exponential function e^x, which is fundamental in calculus due to its unique properties, particularly its derivative. Understanding the behavior of $\frac{e^x - 1}{x}$ near $x = 0$ is essential for comprehending the foundational concepts of limits and derivatives of exponential functions.

To compute $\lim_{x \to 0} \frac{e^x - 1}{x}$, we can employ several methods, including direct substitution, series expansion, and L'Hôpital's Rule. Direct substitution yields an indeterminate form $\frac{0}{0}$, indicating the necessity for further analysis. The Taylor series expansion of e^x around $x = 0$ provides a polynomial approximation that simplifies the limit. Alternatively, L'Hôpital's Rule is applicable to indeterminate forms of type $\frac{0}{0}$ or $\frac{\infty}{\infty}$, allowing differentiation of the numerator and denominator to evaluate the limit.

$$\lim_{x \to 0} \frac{e^x - 1}{x}$$

Applying L'Hôpital's Rule, we differentiate the numerator and the denominator with respect to x:

$$\lim_{x\to 0} \frac{d}{dx}(e^x - 1) \Big/ \frac{d}{dx}(x) = \lim_{x\to 0} \frac{e^x}{1} = e^0 = 1$$

Alternatively, using the Taylor series expansion for e^x around $x = 0$:

$$e^x = 1 + x + \frac{x^2}{2!} + \frac{x^3}{3!} + \cdots$$

Substituting this into the original limit expression:

$$\frac{e^x - 1}{x} = \frac{1 + x + \frac{x^2}{2!} + \frac{x^3}{3!} + \cdots - 1}{x} = \frac{x + \frac{x^2}{2!} + \frac{x^3}{3!} + \cdots}{x} = 1 + \frac{x}{2!} + \frac{x^2}{3!} + \cdots$$

Taking the limit as $x \to 0$:

$$\lim_{x\to 0} \left(1 + \frac{x}{2!} + \frac{x^2}{3!} + \cdots\right) = 1$$

Both methods conclusively show that:

$$\lim_{x\to 0} \frac{e^x - 1}{x} = 1$$

This result is significant as it represents the derivative of e^x at $x = 0$, reinforcing the property that the exponential function is its own derivative. The limit highlights the instantaneous rate of change of e^x at the origin, a concept pivotal in differential calculus.

1.5 Evaluate $\lim_{x\to 0^+} x^x$.

We are tasked with evaluating the limit $\lim_{x\to 0^+} x^x$. This expression involves an indeterminate form because as x approaches 0 from the right, x approaches 0 while x^x appears to involve the form 0^0, which is undefined in standard analysis. To resolve this, we will employ logarithmic transformation and properties of limits to determine the precise value of the limit.

First, consider the function $f(x) = x^x$ for $x > 0$. To simplify the evaluation of the limit, take the natural logarithm of $f(x)$:

$$\ln f(x) = \ln(x^x) = x \ln x.$$

19

Thus, the original limit can be expressed in terms of the natural logarithm:

$$\lim_{x \to 0^+} x^x = \lim_{x \to 0^+} e^{x \ln x} = e^{\lim_{x \to 0^+} x \ln x}.$$

Our next objective is to evaluate $\lim_{x \to 0^+} x \ln x$. As x approaches 0^+, $\ln x$ approaches $-\infty$, and x approaches 0. Therefore, the product $x \ln x$ presents the indeterminate form $0 \cdot (-\infty)$. To resolve this, we can rewrite the expression to a form amenable to L'Hôpital's Rule by expressing it as a ratio:

$$\lim_{x \to 0^+} x \ln x = \lim_{x \to 0^+} \frac{\ln x}{1/x}.$$

In this form, as $x \to 0^+$, $\ln x \to -\infty$ and $1/x \to \infty$, yielding the indeterminate form $\frac{-\infty}{\infty}$. L'Hôpital's Rule states that if the limit of a ratio results in an indeterminate form $\frac{0}{0}$ or $\frac{\infty}{\infty}$, the limit of the ratio of their derivatives may be taken instead, provided this new limit exists.

Applying L'Hôpital's Rule:

$$\lim_{x \to 0^+} \frac{\ln x}{1/x} = \lim_{x \to 0^+} \frac{\frac{d}{dx}(\ln x)}{\frac{d}{dx}(1/x)} = \lim_{x \to 0^+} \frac{\frac{1}{x}}{-\frac{1}{x^2}} = \lim_{x \to 0^+} -x = 0.$$

Therefore, we find that:

$$\lim_{x \to 0^+} x \ln x = 0.$$

Substituting back into our expression for the original limit:

$$\lim_{x \to 0^+} x^x = e^{\lim_{x \to 0^+} x \ln x} = e^0 = 1.$$

Thus, the limit $\lim_{x \to 0^+} x^x$ is equal to 1.

1.6 Prove that $\lim_{x \to 0} \frac{\tan(x)}{x} = 1$.

In this problem, we are tasked with proving that the limit of $\frac{\tan(x)}{x}$ as x approaches 0 is equal to 1. Understanding this limit is fundamental in calculus, particularly in the study of derivatives and trigonometric functions. This limit demonstrates the behavior of the tangent function near the origin and is essential for deriving the derivative of tangent.

To approach this problem, we will utilize fundamental trigonometric identities and known limits. The tangent function can be expressed in terms of sine and cosine:

$$\tan(x) = \frac{\sin(x)}{\cos(x)}$$

Therefore, the expression $\frac{\tan(x)}{x}$ becomes:

$$\frac{\tan(x)}{x} = \frac{\sin(x)}{x \cdot \cos(x)}$$

Our goal is to evaluate the limit:

$$\lim_{x \to 0} \frac{\sin(x)}{x \cdot \cos(x)}$$

We can separate this expression into the product of two limits, provided both limits exist:

$$\lim_{x \to 0} \frac{\sin(x)}{x} \cdot \lim_{x \to 0} \frac{1}{\cos(x)}$$

We know from standard limit results that:

$$\lim_{x \to 0} \frac{\sin(x)}{x} = 1$$

This is a well-established limit in calculus, often proved using the Squeeze Theorem or geometric arguments involving the unit circle.

Next, we evaluate the second limit:

$$\lim_{x \to 0} \frac{1}{\cos(x)}$$

Since $\cos(0) = 1$, and the cosine function is continuous at $x = 0$, it follows that:

$$\lim_{x \to 0} \cos(x) = \cos(0) = 1$$

Therefore, the limit simplifies to:

$$\lim_{x \to 0} \frac{1}{\cos(x)} = \frac{1}{\lim_{x \to 0} \cos(x)} = \frac{1}{1} = 1$$

Combining the two limits, we have:

$$\lim_{x \to 0} \frac{\tan(x)}{x} = \lim_{x \to 0} \frac{\sin(x)}{x \cdot \cos(x)} = \left(\lim_{x \to 0} \frac{\sin(x)}{x} \right) \cdot \left(\lim_{x \to 0} \frac{1}{\cos(x)} \right) = 1 \cdot 1 = 1$$

Thus, we have rigorously demonstrated that:

$$\lim_{x \to 0} \frac{\tan(x)}{x} = 1$$

This result not only confirms the behavior of the tangent function near zero but also serves as a foundational limit in the study of derivatives involving trigonometric functions.

1.7 Determine the points of discontinuity of a piecewise function.

Consider the function $f : \mathbb{R} \to \mathbb{R}$ defined by

$$f(x) = \begin{cases} \frac{x^2 - 4}{x - 2} & \text{if } x \neq 2, \\ 3 & \text{if } x = 2. \end{cases}$$

To determine the points of discontinuity of $f(x)$, we analyze the behavior of the function at all points in its domain, with particular attention to the points where the definition of $f(x)$ changes, which in this case is $x = 2$.

- A function $f(x)$ is said to be continuous at a point $x = a$ if the following three conditions are satisfied:

    - $f(a)$ is defined.
    - $\lim_{x \to a} f(x)$ exists.
    - $\lim_{x \to a} f(x) = f(a)$.

- If any of these conditions fail, $f(x)$ is discontinuous at $x = a$.

- First, we identify the points where the function's definition changes or where the function may be undefined. In the given piecewise function, the only point of interest is $x = 2$.

- For $x \neq 2$, the function simplifies as follows:

$$f(x) = \frac{x^2 - 4}{x - 2} = \frac{(x - 2)(x + 2)}{x - 2} = x + 2, \quad \text{for } x \neq 2.$$

- This simplification is valid for all x except $x = 2$, where the original expression $\frac{x^2 - 4}{x - 2}$ is undefined.

22

- We now evaluate the three conditions for continuity at $x = 2$:

  - **Function is defined at $x = 2$:**
    $f(2) = 3$ is defined.

  - **Limit as x approaches 2 exists:**
    Since $f(x) = x + 2$ for $x \neq 2$, we compute

    $$\lim_{x \to 2} f(x) = \lim_{x \to 2} (x + 2) = 4.$$

    The limit exists and is equal to 4.

  - **Limit equals function value at $x = 2$:**
    $\lim_{x \to 2} f(x) = 4$ and $f(2) = 3$. Since $4 \neq 3$, this condition is not satisfied.

- Since the third condition for continuity is not met at $x = 2$, the function $f(x)$ is discontinuous at this point.

- Examining the entire domain of $f(x)$, the only point of discontinuity is at $x = 2$. For all other real numbers $x \neq 2$, $f(x)$ simplifies to $x + 2$, which is a continuous function. Therefore, $x = 2$ is the sole point where $f(x)$ is discontinuous.

To determine the points of discontinuity of a piecewise function:

- Identify the points where the function's definition changes or where it may be undefined.

- For each such point $x = a$, verify the three conditions for continuity:

  - $f(a)$ is defined.
  - $\lim_{x \to a} f(x)$ exists.
  - $\lim_{x \to a} f(x) = f(a)$.

- If any condition fails, $f(x)$ is discontinuous at $x = a$.

Applying this method to the given function, we conclude that $f(x)$ is discontinuous only at $x = 2$.

- The approach demonstrated can be generalized to any piecewise function. When dealing with piecewise functions:

- Examine each point where the function's definition changes.
- Simplify the function in regions where it is defined differently.
- Compute the limits from the left and right at each critical point.
- Compare the limits to the function's value at those points to ascertain continuity.

This systematic method ensures a comprehensive analysis of discontinuities in piecewise functions.

1.8 Investigate the limit $\lim_{x \to 0} \frac{x - \sin(x)}{x^3}$.

The limit $\lim_{x \to 0} \frac{x - \sin(x)}{x^3}$ involves determining the behavior of the function $\frac{x - \sin(x)}{x^3}$ as x approaches zero. Evaluating such limits is fundamental in calculus, particularly in understanding the finer properties of functions near specific points. This limit requires an analysis of the difference between the linear function x and the trigonometric function $\sin(x)$, normalized by x^3. The challenge lies in accurately capturing the rate at which $x - \sin(x)$ approaches zero relative to x^3.

To evaluate this limit, we can utilize the Taylor series expansion of $\sin(x)$ around $x = 0$. The Maclaurin series for $\sin(x)$ is given by:

$$\sin(x) = x - \frac{x^3}{6} + \frac{x^5}{120} - \frac{x^7}{5040} + \cdots$$

This expansion provides an approximation of $\sin(x)$ in terms of increasingly higher powers of x. By subtracting $\sin(x)$ from x, we obtain:

$$x - \sin(x) = x - \left(x - \frac{x^3}{6} + \frac{x^5}{120} - \cdots \right)$$

which simplifies to:

$$x - \sin(x) = \frac{x^3}{6} - \frac{x^5}{120} + \cdots$$

Notice that the leading term in this difference is $\frac{x^3}{6}$, with higher-order terms involving higher powers of x. Dividing this expression by x^3 yields:

$$\frac{x - \sin(x)}{x^3} = \frac{\frac{x^3}{6} - \frac{x^5}{120} + \cdots}{x^3} = \frac{1}{6} - \frac{x^2}{120} + \cdots$$

As x approaches zero, the higher-order terms containing x^2, x^4, etc., become negligible. Therefore, the limit simplifies to:

$$\lim_{x \to 0} \frac{x - \sin(x)}{x^3} = \frac{1}{6}$$

This result indicates that as x approaches zero, the ratio $\frac{x-\sin(x)}{x^3}$ approaches $\frac{1}{6}$. The use of the Taylor series expansion allows for a precise calculation by systematically accounting for the contributions of each power of x in the vicinity of zero. This method not only provides the exact value of the limit but also offers deeper insight into the local behavior of the sine function compared to a linear approximation.

The limit $\lim_{x \to 0} \frac{x-\sin(x)}{x^3}$ is equal to $\frac{1}{6}$. This conclusion is reached by expanding $\sin(x)$ into its Maclaurin series, simplifying the expression, and observing the behavior of the resulting terms as x approaches zero.

1.9 Evaluate $\lim_{x \to \infty} \frac{\ln(x)}{x}$.

In this problem, we are tasked with evaluating the limit of the ratio of the natural logarithm of x to x as x approaches infinity. Mathematically, this is expressed as:

$$\lim_{x \to \infty} \frac{\ln(x)}{x}$$

Understanding the behavior of this limit requires an analysis of how the functions $\ln(x)$ and x grow as x becomes large. The natural logarithm function $\ln(x)$ grows without bound as x increases, but it does so at a much slower rate compared to the linear function x. The key question is to quantify this difference in growth rates and determine the limiting value of their ratio.

To evaluate the limit, we observe that as $x \to \infty$, both $\ln(x)$ and x approach infinity, leading to an indeterminate form of type $\frac{\infty}{\infty}$. In such cases, L'Hôpital's Rule is a powerful tool that can be applied to resolve the indeterminacy. L'Hôpital's Rule states that if the limit $\lim_{x \to c} \frac{f(x)}{g(x)}$ yields an indeterminate form $\frac{0}{0}$ or $\frac{\infty}{\infty}$, then:

$$\lim_{x \to c} \frac{f(x)}{g(x)} = \lim_{x \to c} \frac{f'(x)}{g'(x)}$$

provided the limit on the right-hand side exists.

Applying L'Hôpital's Rule to our limit, we first differentiate the numerator and the denominator with respect to x:

25

- Differentiate the numerator:

$$\frac{d}{dx}\ln(x) = \frac{1}{x}$$

- Differentiate the denominator:

$$\frac{d}{dx}x = 1$$

Now, we form the new limit using these derivatives:

$$\lim_{x\to\infty}\frac{\frac{1}{x}}{1} = \lim_{x\to\infty}\frac{1}{x}$$

As x approaches infinity, the expression $\frac{1}{x}$ approaches zero. Therefore, the limit simplifies to:

$$\lim_{x\to\infty}\frac{1}{x} = 0$$

This result indicates that the natural logarithm function $\ln(x)$ grows significantly slower than the linear function x as x becomes large. Consequently, the ratio $\frac{\ln(x)}{x}$ diminishes towards zero in the limit as x approaches infinity. This conclusion aligns with our initial intuition about the relative growth rates of logarithmic and linear functions.

$$\boxed{0}$$

1.10 Find the horizontal asymptotes of the function $f(x) = \frac{2x^2}{x^2+1}$.

Determining the horizontal asymptotes of a rational function involves analyzing the behavior of the function as the independent variable approaches positive and negative infinity. Horizontal asymptotes represent the end behavior of the function, indicating the values that the function approaches but does not necessarily reach as x becomes very large in magnitude. For the given function $f(x) = \frac{2x^2}{x^2+1}$, both the numerator and the denominator are polynomials of degree 2. The degrees of the numerator and the denominator play a crucial role in determining the horizontal asymptotes. When the degrees of the numerator and the denominator are equal, the horizontal asymptote can be found by taking the ratio of the leading coefficients of the numerator and the denominator.

Therefore, the analysis involves identifying the leading coefficients and examining the limit of the function as x approaches positive and negative infinity to ascertain the presence and value of any horizontal asymptotes.

To find the horizontal asymptotes of the function $f(x) = \frac{2x^2}{x^2+1}$, we begin by identifying the degrees of the numerator and the denominator. The numerator, $2x^2$, is a polynomial of degree 2, and the denominator, $x^2 + 1$, is also a polynomial of degree 2. Since the degrees of the numerator and the denominator are equal, the horizontal asymptote is determined by the ratio of the leading coefficients of these polynomials.

The leading coefficient of the numerator $2x^2$ is 2, and the leading coefficient of the denominator $x^2 + 1$ is 1. Therefore, the horizontal asymptote is given by the ratio of these leading coefficients:

$$y = \frac{2}{1} = 2$$

This indicates that as x approaches positive or negative infinity, the function $f(x)$ approaches the value of 2.

To confirm this result, we evaluate the limit of $f(x)$ as x approaches infinity and negative infinity:

$$\lim_{x \to \infty} \frac{2x^2}{x^2 + 1} = \lim_{x \to \infty} \frac{2}{1 + \frac{1}{x^2}} = \frac{2}{1 + 0} = 2$$

$$\lim_{x \to -\infty} \frac{2x^2}{x^2 + 1} = \lim_{x \to -\infty} \frac{2}{1 + \frac{1}{x^2}} = \frac{2}{1 + 0} = 2$$

Both limits confirm that $y = 2$ is the horizontal asymptote of the function $f(x)$, as the function approaches this value from both directions as x becomes very large in magnitude.

$$\boxed{y = 2}$$

Chapter 2

Differentiation Techniques

This chapter delves into a diverse set of differentiation techniques, providing a comprehensive set of problems aimed at enhancing proficiency in calculating derivatives. The focus lies on applying a variety of methods such as the product, quotient, and chain rules, along with implicit differentiation and higher-order derivatives. Through tackling these challenging problems, readers will build a strong foundation in differentiation, essential for advanced calculus topics and applications. Each problem is supplemented with detailed solutions to help clarify complex concepts and solidify understanding.

2.1 Differentiate $y = x^x$.

To determine the derivative of the function $y = x^x$ with respect to x, we encounter a unique scenario where both the base and the exponent are variables dependent on x. This complexity necessitates the application of logarithmic differentiation, a powerful technique useful for differentiating functions where standard differentiation rules, such as the power rule, are insufficient.

Problem Explanation:

The function $y = x^x$ presents a case where the base x and the exponent x both vary with x. Unlike standard power functions of the form $y = x^n$, where n is a constant, the presence of variable exponents requires a different approach. Direct application of the power rule is not feasible here because both components of the function are functions of x. Consequently, logarithmic differentiation becomes the appropriate method to find $\frac{dy}{dx}$.

Solution:

We proceed with the differentiation using logarithmic differentiation as follows:

- **Apply the Natural Logarithm:**

  Start by taking the natural logarithm of both sides of the equation to simplify the differentiation process.

  $$\ln y = \ln(x^x)$$

- **Simplify Using Logarithmic Identities:**

  Utilize the logarithmic identity $\ln(a^b) = b \ln a$ to bring the exponent down as a coefficient.

  $$\ln y = x \ln x$$

- **Differentiate Both Sides Implicitly:**

  Differentiate both sides of the equation with respect to x. This step involves applying the chain rule on the left-hand side and the product rule on the right-hand side.

  $$\frac{d}{dx}(\ln y) = \frac{d}{dx}(x \ln x)$$

- **Apply the Chain Rule to the Left Side:**

  The derivative of $\ln y$ with respect to y is $\frac{1}{y}$, and by the chain rule, we multiply by $\frac{dy}{dx}$.

  $$\frac{1}{y} \cdot \frac{dy}{dx} = \frac{d}{dx}(x \ln x)$$

- **Apply the Product Rule to the Right Side:**

  To differentiate $x \ln x$, where both x and $\ln x$ are functions of x, we use the product rule. Letting $u = x$ and $v = \ln x$, the product rule states that $\frac{d}{dx}(uv) = u'v + uv'$.

  – $u' = \frac{d}{dx} x = 1$
  – $v' = \frac{d}{dx} \ln x = \frac{1}{x}$

30

Applying the product rule:

$$\frac{d}{dx}(x \ln x) = (1)(\ln x) + (x)\left(\frac{1}{x}\right) = \ln x + 1$$

- **Equate the Derivatives:**

  Substituting the derivative of the right side back into the equation, we have:

$$\frac{1}{y} \cdot \frac{dy}{dx} = \ln x + 1$$

- **Solve for $\frac{dy}{dx}$:**

  Multiply both sides of the equation by y to isolate $\frac{dy}{dx}$:

$$\frac{dy}{dx} = y(\ln x + 1)$$

- **Substitute Back $y = x^x$:**

  Finally, replace y with x^x to express the derivative entirely in terms of x:

$$\frac{dy}{dx} = x^x(1 + \ln x)$$

Conclusion:

The derivative of the function $y = x^x$ with respect to x is thus:

$$\boxed{\frac{dy}{dx} = x^x(1 + \ln x)}$$

This result showcases the efficacy of logarithmic differentiation in handling functions where both the base and the exponent are variable, extending the repertoire of differentiation techniques necessary for tackling a broad spectrum of calculus problems.

2.2 Find $\frac{dy}{dx}$ if $y = \ln(\sin(x))$.

The problem requires finding the derivative of the function $y = \ln(\sin(x))$ with respect to x. To determine $\frac{dy}{dx}$, we apply the rules of differentiation, specifically the chain rule. The chain rule is essential when differentiating composite functions, where one function is nested within another. In this case, $\ln(\sin(x))$ is a composition of the natural logarithm function and the sine function.

Begin by identifying the outer function and the inner function in the composition. Here, the outer function is $\ln(u)$, where $u = \sin(x)$. The derivative of the natural logarithm function $\ln(u)$ with respect to u is $\frac{1}{u}$. The inner function is $\sin(x)$, and its derivative with respect to x is $\cos(x)$.

Applying the chain rule, which states that the derivative of a composite function $y = f(g(x))$ is $y' = f'(g(x)) \cdot g'(x)$, we proceed as follows:

$$\frac{dy}{dx} = \frac{d}{dx}\left[\ln(\sin(x))\right] = \frac{d}{du}\left[\ln(u)\right] \cdot \frac{d}{dx}\left[\sin(x)\right]$$

Substituting the derivatives of the outer and inner functions:

$$\frac{dy}{dx} = \frac{1}{u} \cdot \cos(x) = \frac{1}{\sin(x)} \cdot \cos(x)$$

Simplifying the expression:

$$\frac{dy}{dx} = \frac{\cos(x)}{\sin(x)}$$

Recognizing that $\frac{\cos(x)}{\sin(x)}$ is the cotangent function:

$$\frac{dy}{dx} = \cot(x)$$

Thus, the derivative of $y = \ln(\sin(x))$ with respect to x is $\cot(x)$.

2.3 Compute the derivative of $y = e^{x^2}$.

To determine the derivative of the function $y = e^{x^2}$ with respect to x, we must apply the principles of differentiation, particularly the chain rule. The chain rule is essential when dealing with composite functions, where one function is nested within another. In this case, the exponential function e^u is composed with the quadratic function $u = x^2$. Understanding each component and how they interact is crucial for accurate differentiation.

The function $y = e^{x^2}$ can be viewed as a composition of two functions:

$$y = e^u \quad \text{where} \quad u = x^2$$

The chain rule states that the derivative of a composite function $y = f(g(x))$ is the derivative of the outer function evaluated at the inner function multiplied by the derivative of the inner function:

$$\frac{dy}{dx} = \frac{df}{dg} \cdot \frac{dg}{dx}$$

Applying this to our function, we identify $f(u) = e^u$ and $g(x) = x^2$.

First, we compute the derivative of the outer function with respect to the inner function u:

$$\frac{df}{du} = \frac{d}{du} e^u = e^u$$

Next, we compute the derivative of the inner function u with respect to x:

$$\frac{du}{dx} = \frac{d}{dx} x^2 = 2x$$

Finally, applying the chain rule, we multiply these two derivatives to obtain the derivative of y with respect to x:

$$\frac{dy}{dx} = \frac{df}{du} \cdot \frac{du}{dx} = e^u \cdot 2x = 2x e^{x^2}$$

Thus, the derivative of $y = e^{x^2}$ with respect to x is:

$$\frac{dy}{dx} = 2x e^{x^2}$$

This result demonstrates the application of the chain rule to a composite function, where the exponential function's rapid growth is modulated by the quadratic term in the exponent. The derivative $\frac{dy}{dx} = 2x e^{x^2}$ effectively captures how changes in x influence the rate of growth of the function y.

2.4 Differentiate $y = \arcsin(x)$.

To differentiate the function $y = \arcsin(x)$, we seek to find the derivative $\frac{dy}{dx}$. The arcsine function, denoted as $\arcsin(x)$, is the inverse of the sine function restricted to the interval $\left[-\frac{\pi}{2}, \frac{\pi}{2}\right]$. Differentiating inverse trigonometric functions typically involves implicit differentiation or the use of inverse function derivatives. Here, we will employ both methods to derive the derivative of $y = \arcsin(x)$.

$$y = \arcsin(x)$$

Method 1: Using the Inverse Function Derivative Formula

The inverse function derivative formula states that if $y = f^{-1}(x)$, then

$$\frac{dy}{dx} = \frac{1}{f'(y)}$$

In our case, $f(y) = \sin(y)$, so $f^{-1}(x) = \arcsin(x)$. Therefore, we have:

$$\frac{dy}{dx} = \frac{1}{\frac{d}{dy}\sin(y)} = \frac{1}{\cos(y)}$$

To express $\frac{dy}{dx}$ in terms of x, we recall the Pythagorean identity:

$$\sin^2(y) + \cos^2(y) = 1$$

Since $y = \arcsin(x)$, we have $\sin(y) = x$. Substituting into the identity:

$$x^2 + \cos^2(y) = 1 \implies \cos^2(y) = 1 - x^2$$

Taking the positive square root (since $\cos(y)$ is positive in the range $\left[-\frac{\pi}{2}, \frac{\pi}{2}\right]$):

$$\cos(y) = \sqrt{1 - x^2}$$

Substituting back into the derivative:

$$\frac{dy}{dx} = \frac{1}{\sqrt{1 - x^2}}$$

Method 2: Implicit Differentiation

Alternatively, we can use implicit differentiation to find $\frac{dy}{dx}$. Starting with the original equation:

34

$$y = \arcsin(x)$$

Taking the sine of both sides to eliminate the arcsine function:

$$\sin(y) = x$$

Differentiate both sides with respect to x:

$$\frac{d}{dx}\sin(y) = \frac{d}{dx}x$$

Applying the chain rule to the left side:

$$\cos(y) \cdot \frac{dy}{dx} = 1$$

Solving for $\frac{dy}{dx}$:

$$\frac{dy}{dx} = \frac{1}{\cos(y)}$$

As in Method 1, we use the Pythagorean identity to express $\cos(y)$ in terms of x:

$$\cos(y) = \sqrt{1 - x^2}$$

Substituting back:

$$\frac{dy}{dx} = \frac{1}{\sqrt{1 - x^2}}$$

Both methods yield the same result for the derivative of $y = \arcsin(x)$:

$$\frac{dy}{dx} = \frac{1}{\sqrt{1 - x^2}}$$

This derivative is valid for x in the open interval $(-1, 1)$, where the function $y = \arcsin(x)$ is differentiable.

2.5 Find the second derivative of $y = \tan(x)$.

In this problem, we are tasked with determining the second derivative of the function $y = \tan(x)$. The second derivative of a function provides information about the concavity of the function and the acceleration of its rate of change. Calculating the second derivative involves first finding the first derivative, and then differentiating that result once more with respect to x. This process requires a solid understanding of differentiation rules, particularly those applicable to trigonometric functions.

To begin, recall that the first derivative of $y = \tan(x)$ represents the rate at which y changes with respect to x. The derivative of the tangent function is a standard result in calculus, which we will utilize in our calculations.

y = \tan(x)

First, we compute the first derivative $\frac{dy}{dx}$:

$$\frac{dy}{dx} = \frac{d}{dx}\tan(x)$$

The derivative of $\tan(x)$ with respect to x is $\sec^2(x)$. This is a fundamental derivative in calculus involving trigonometric functions.

$$\frac{dy}{dx} = \sec^2(x)$$

Next, we proceed to find the second derivative $\frac{d^2y}{dx^2}$, which involves differentiating $\frac{dy}{dx}$ with respect to x:

$$\frac{d^2y}{dx^2} = \frac{d}{dx}\left(\sec^2(x)\right)$$

To differentiate $\sec^2(x)$, we apply the chain rule. Recall that $\sec(x) = \frac{1}{\cos(x)}$, and the derivative of $\sec(x)$ with respect to x is $\sec(x)\tan(x)$. However, since we have $\sec^2(x)$, we treat it as a composite function $[\sec(x)]^2$.

Applying the chain rule:

$$\frac{d}{dx}\left(\sec^2(x)\right) = 2\sec(x) \cdot \frac{d}{dx}\sec(x)$$

36

We already know that $\frac{d}{dx}\sec(x) = \sec(x)\tan(x)$. Substituting this into the equation:

$$\frac{d}{dx}\left(\sec^2(x)\right) = 2\sec(x)\cdot\sec(x)\tan(x)$$

Simplifying the expression:

$$\frac{d}{dx}\left(\sec^2(x)\right) = 2\sec^2(x)\tan(x)$$

Therefore, the second derivative of $y = \tan(x)$ with respect to x is:

$$\frac{d^2y}{dx^2} = 2\sec^2(x)\tan(x)$$

This result illustrates how the second derivative of a trigonometric function can be derived using standard differentiation techniques, including the chain rule and the derivatives of basic trigonometric identities. Understanding this process is fundamental for analyzing the curvature and concavity of trigonometric functions in various applications.

2.6 Compute $\frac{d}{dx}\left(\ln(x^2+1)\right)$.

To compute the derivative of the function $y = \ln(x^2+1)$ with respect to x, we utilize the chain rule, a fundamental technique in differentiation for handling composite functions. The chain rule states that if a function y can be expressed as the composition of two functions f and g, such that $y = f(g(x))$, then the derivative of y with respect to x is given by:

$$\frac{dy}{dx} = f'(g(x))\cdot g'(x)$$

In the context of $y = \ln(x^2+1)$, we identify the outer function $f(u) = \ln(u)$ and the inner function $g(x) = x^2 + 1$.

The derivative of the outer function $f(u)$ with respect to u is:

$$f'(u) = \frac{d}{du}\ln(u) = \frac{1}{u}$$

The derivative of the inner function $g(x)$ with respect to x is:

$$g'(x) = \frac{d}{dx}(x^2+1) = 2x$$

Applying the chain rule, we multiply these derivatives:

$$\frac{dy}{dx} = f'(g(x)) \cdot g'(x) = \frac{1}{x^2+1} \cdot 2x = \frac{2x}{x^2+1}$$

Therefore, the derivative of $y = \ln(x^2+1)$ with respect to x is:

$$\frac{dy}{dx} = \frac{2x}{x^2+1}$$

This result illustrates the application of the chain rule in differentiating logarithmic functions involving polynomial expressions. The derivative provides the rate at which y changes with respect to x, which is essential in various applications such as optimization, curve sketching, and understanding the behavior of functions.

2.7 Differentiate implicitly the equation $x^2 + y^2 = 1$.

In this problem, we are tasked with finding the derivative $\frac{dy}{dx}$ of the implicitly defined function y in terms of x from the equation $x^2 + y^2 = 1$. Implicit differentiation is necessary here because the equation defines y implicitly as a function of x, rather than providing y explicitly in terms of x.

To perform implicit differentiation, we differentiate both sides of the equation with respect to x, treating y as a function of x. This means that when differentiating terms involving y, we must apply the chain rule, recognizing that y depends on x.

Starting with the original equation:

$$x^2 + y^2 = 1$$

Differentiate both sides with respect to x:

$$\frac{d}{dx}\left(x^2\right) + \frac{d}{dx}\left(y^2\right) = \frac{d}{dx}(1)$$

Compute each derivative separately:

$$\frac{d}{dx}\left(x^2\right) = 2x$$

$$\frac{d}{dx}\left(y^2\right) = 2y \cdot \frac{dy}{dx}$$

$$\frac{d}{dx}(1) = 0$$

Substituting these results back into the differentiated equation:

$$2x + 2y \cdot \frac{dy}{dx} = 0$$

Our goal is to solve for $\frac{dy}{dx}$. To isolate $\frac{dy}{dx}$, perform the following algebraic steps:

- Subtract $2x$ from both sides:

$$2y \cdot \frac{dy}{dx} = -2x$$

- Divide both sides by $2y$:

$$\frac{dy}{dx} = \frac{-2x}{2y}$$

- Simplify the expression by canceling the common factor of 2:

$$\frac{dy}{dx} = \frac{-x}{y}$$

Thus, the derivative of y with respect to x is:

$$\frac{dy}{dx} = -\frac{x}{y}$$

This result indicates that the slope of the tangent line to the curve defined by $x^2 + y^2 = 1$ at any point (x, y) is equal to the negative ratio of the x-coordinate to the y-coordinate at that point. This derivative is essential in understanding the behavior of the function y in relation to x without requiring an explicit expression for y in terms of x.

2.8 Find $\frac{dy}{dx}$ for $y = x^x$.

Problem Explanation

The function $y = x^x$ presents a scenario where both the base and the exponent are functions of the variable x. Traditional differentiation rules, such as the power rule, are not directly applicable to this form because they typically handle cases where either the base or the exponent is constant, but not both. To differentiate $y = x^x$, we utilize logarithmic differentiation, a powerful technique that simplifies the differentiation of functions where both the base and the exponent involve the variable.

Solution

To determine $\frac{dy}{dx}$ for $y = x^x$, follow these detailed steps:

- Apply the Natural Logarithm to Both Sides:

  Start by taking the natural logarithm of both sides of the equation to exploit the properties of logarithms in simplifying the expression:

$$\ln y = \ln(x^x)$$

- Simplify Using Logarithmic Identities:

  Utilize the logarithmic identity $\ln(a^b) = b \ln a$ to simplify the right-hand side:

$$\ln y = x \ln x$$

- Differentiate Both Sides with Respect to x:

  Differentiate both sides of the equation with respect to x. This step requires the application of implicit differentiation on the left side and the product rule on the right side:

$$\frac{d}{dx}(\ln y) = \frac{d}{dx}(x \ln x)$$

- Differentiate the Left Side Using Implicit Differentiation:

  The left side involves differentiating $\ln y$ with respect to x. Applying the chain rule yields:

$$\frac{1}{y} \cdot \frac{dy}{dx} = \frac{d}{dx}(x \ln x)$$

- Differentiate the Right Side Using the Product Rule:

  The right side is the derivative of the product $x \ln x$. Apply the product rule, where if $u = x$ and $v = \ln x$, then $\frac{d}{dx}(uv) = u\frac{dv}{dx} + v\frac{du}{dx}$:

$$\frac{d}{dx}(x \ln x) = x \cdot \frac{1}{x} + \ln x \cdot 1 = 1 + \ln x$$

- Equate the Derivatives:

  Set the derivatives from both sides equal to each other:

  $$\frac{1}{y} \cdot \frac{dy}{dx} = 1 + \ln x$$

- Solve for $\frac{dy}{dx}$:

  Isolate $\frac{dy}{dx}$ by multiplying both sides of the equation by y:

  $$\frac{dy}{dx} = y \cdot (1 + \ln x)$$

- Substitute $y = x^x$ Back into the Equation:

  Replace y with the original expression x^x:

  $$\frac{dy}{dx} = x^x \cdot (1 + \ln x)$$

- Final Expression for the Derivative:

  The derivative of $y = x^x$ with respect to x is therefore:

  $$\frac{dy}{dx} = x^x(1 + \ln x)$$

Through the application of logarithmic differentiation, we effectively handled the complexity of differentiating a function where both the base and the exponent are variable. The final derivative, $\frac{dy}{dx} = x^x(1 + \ln x)$, succinctly captures the rate of change of y with respect to x for the given function.

2.9 Differentiate $y = \frac{x}{\sqrt{x^2+1}}$.

To differentiate the function $y = \frac{x}{\sqrt{x^2+1}}$, we will employ the quotient rule combined with the chain rule. This function is a ratio of two differentiable functions: the numerator $u(x) = x$ and the denominator $v(x) = \sqrt{x^2 + 1}$. The quotient rule is appropriate here because it provides a systematic method for differentiating a quotient of two functions.

The quotient rule states that if $y = \frac{u(x)}{v(x)}$, then the derivative y' is given by:

$$y' = \frac{u'(x)v(x) - u(x)v'(x)}{[v(x)]^2}$$

41

Applying this rule requires finding the derivatives of both $u(x)$ and $v(x)$.

First, compute the derivative of the numerator $u(x) = x$:

$$u'(x) = \frac{d}{dx}[x] = 1$$

Next, compute the derivative of the denominator $v(x) = \sqrt{x^2 + 1}$. To differentiate $v(x)$, we can rewrite it using a fractional exponent:

$$v(x) = (x^2 + 1)^{1/2}$$

Applying the chain rule, which states that the derivative of $(g(x))^n$ is $n(g(x))^{n-1} \cdot g'(x)$, we get:

$$v'(x) = \frac{1}{2}(x^2 + 1)^{-1/2} \cdot \frac{d}{dx}[x^2 + 1] = \frac{1}{2}(x^2 + 1)^{-1/2} \cdot 2x = \frac{x}{\sqrt{x^2 + 1}}$$

Now, substitute $u(x)$, $u'(x)$, $v(x)$, and $v'(x)$ into the quotient rule formula:

$$y' = \frac{1 \cdot \sqrt{x^2 + 1} - x \cdot \frac{x}{\sqrt{x^2+1}}}{(\sqrt{x^2 + 1})^2}$$

Simplify each term in the numerator:

$$1 \cdot \sqrt{x^2 + 1} = \sqrt{x^2 + 1}$$

$$x \cdot \frac{x}{\sqrt{x^2 + 1}} = \frac{x^2}{\sqrt{x^2 + 1}}$$

Therefore, the numerator becomes:

$$\sqrt{x^2 + 1} - \frac{x^2}{\sqrt{x^2 + 1}} = \frac{(x^2 + 1) - x^2}{\sqrt{x^2 + 1}} = \frac{1}{\sqrt{x^2 + 1}}$$

The denominator simplifies as follows:

$$(\sqrt{x^2 + 1})^2 = x^2 + 1$$

Substituting back into the expression for y':

$$y' = \frac{\frac{1}{\sqrt{x^2+1}}}{x^2 + 1} = \frac{1}{(x^2 + 1)^{3/2}}$$

Thus, the derivative of the function $y = \frac{x}{\sqrt{x^2+1}}$ with respect to x is:

$$\boxed{\frac{dy}{dx} = \frac{1}{(x^2 + 1)^{3/2}}}$$

2.10 Compute the derivative of $y = \ln|\sec(x)|$.

To determine the derivative of the function $y = \ln|\sec(x)|$, we begin by ana-
lyzing the composition of the function and identifying the necessary differentiation techniques. The function involves the natural logarithm of the absolute
value of the secant function, which necessitates the application of the chain
rule and an understanding of the properties of absolute values and trigonometric functions.

First, recall that the secant function is defined as $\sec(x) = \frac{1}{\cos(x)}$. The absolute
value ensures that the argument of the logarithm is always positive, which is
essential since the logarithm function is only defined for positive real numbers.
Specifically, $|\sec(x)| = \frac{1}{|\cos(x)|}$, ensuring the argument of the logarithm re-
mains positive regardless of the value of x within its domain.

The function can thus be expressed as:

$$y = \ln|\sec(x)| = \ln\left(\frac{1}{|\cos(x)|}\right) = -\ln|\cos(x)|$$

This reformulation simplifies the differentiation process by expressing the function in terms of $\cos(x)$, a more familiar trigonometric function.

To find the derivative $\frac{dy}{dx}$, we apply the chain rule, which states that the derivative of a composite function $f(g(x))$ is $f'(g(x)) \cdot g'(x)$. Applying this to
$y = -\ln|\cos(x)|$, we proceed as follows:

$$\frac{dy}{dx} = -\frac{d}{dx}\left[\ln|\cos(x)|\right]$$

The derivative of $\ln|u|$ with respect to u is $\frac{1}{u}$, and then we multiply by the
derivative of u with respect to x. Therefore:

$$\frac{dy}{dx} = -\left(\frac{1}{|\cos(x)|} \cdot \frac{d}{dx}|\cos(x)|\right)$$

Next, we differentiate $|\cos(x)|$ with respect to x. The derivative of $|\cos(x)|$
depends on the sign of $\cos(x)$:

$$\frac{d}{dx}|\cos(x)| = \frac{d}{dx}\cos(x) \cdot \text{sgn}(\cos(x)) = -\sin(x) \cdot \text{sgn}(\cos(x))$$

Here, $\text{sgn}(\cos(x))$ represents the sign function, which equals 1 when $\cos(x) >$
0 and -1 when $\cos(x) < 0$. However, since $\frac{1}{|\cos(x)|} \cdot \text{sgn}(\cos(x)) = \frac{\cos(x)}{|\cos(x)|} =$
± 1, the expression simplifies significantly:

$$\frac{dy}{dx} = -\left(\frac{1}{|\cos(x)|} \cdot (-\sin(x) \cdot \text{sgn}(\cos(x)))\right) = \frac{\sin(x)}{\cos(x)} = \tan(x)$$

Therefore, the derivative of $y = \ln|\sec(x)|$ is:

$$\frac{dy}{dx} = \tan(x)$$

This result indicates that the rate of change of $\ln|\sec(x)|$ with respect to x is equal to the tangent of x. The computation leverages fundamental differentiation rules, including the chain rule and the properties of absolute values and trigonometric functions, to arrive at a concise and exact expression for the derivative.

Chapter 3

Applications of Derivatives

This chapter explores the practical utility of derivatives through a variety of applications in calculus. It encompasses problems that address topics such as optimization, related rates, curve sketching, and motion analysis. The objective is to demonstrate how the theoretical principles of derivatives translate into real-world problem-solving scenarios. Readers will engage in analytical reasoning and practical computations, supported by comprehensive solutions that elucidate the steps and thought processes involved in each application. This approach aims to reinforce the relevance and applicability of derivatives in diverse contexts.

3.1 Find the critical points of $f(x) = x^3 - 3x^2 + 4$.

Critical points of a function are points on its graph where the first derivative is zero or undefined. These points are significant as they often correspond to local maxima, local minima, or points of inflection, thereby providing valuable information about the behavior of the function. Identifying critical points is essential in optimization problems and in sketching the graph of the function. In this problem, we aim to determine the critical points of the cubic function $f(x) = x^3 - 3x^2 + 4$.

To find the critical points of $f(x)$, we follow these steps:

1. Compute the first derivative of $f(x)$: The first derivative $f'(x)$ represents the slope of the tangent line to the graph of $f(x)$ at any point x. Calculating $f'(x)$ will allow us to identify where the slope is zero or

undefined.

2. Set the first derivative equal to zero and solve for x: The values of x that satisfy $f'(x) = 0$ are potential critical points, as they indicate points where the function has horizontal tangent lines.

3. Determine the corresponding y-values: For each x value obtained, substitute it back into the original function $f(x)$ to find the corresponding y-value, thereby identifying the coordinates of the critical points.

4. Verify if the derivative is defined at these points: Ensure that the first derivative exists at the critical points. Points where the derivative does not exist could also be critical points, but in this case, since $f(x)$ is a polynomial, the derivative exists everywhere.

Let's apply these steps to the given function $f(x) = x^3 - 3x^2 + 4$.

Step 1: Compute the first derivative The first derivative of $f(x)$ with respect to x is obtained by differentiating each term of the function separately:

$$f'(x) = \frac{d}{dx}\left(x^3\right) - \frac{d}{dx}\left(3x^2\right) + \frac{d}{dx}\left(4\right)$$

Calculating each derivative:

$$\frac{d}{dx}\left(x^3\right) = 3x^2$$

$$\frac{d}{dx}\left(3x^2\right) = 6x$$

$$\frac{d}{dx}\left(4\right) = 0$$

Combining these results:

$$f'(x) = 3x^2 - 6x$$

Step 2: Set the first derivative equal to zero and solve for x Set $f'(x) = 0$:

$$3x^2 - 6x = 0$$

Factor the equation:

$$3x(x - 2) = 0$$

Set each factor equal to zero and solve for x:

$$3x = 0 \quad \Rightarrow \quad x = 0$$

$$x - 2 = 0 \quad \Rightarrow \quad x = 2$$

Thus, the potential critical points occur at $x = 0$ and $x = 2$.

Step 3: Determine the corresponding y-values Substitute each x-value back into the original function $f(x)$ to find the corresponding y-values.

For $x = 0$:

$$f(0) = (0)^3 - 3(0)^2 + 4 = 0 - 0 + 4 = 4$$

Thus, the first critical point is $(0, 4)$.

For $x = 2$:

$$f(2) = (2)^3 - 3(2)^2 + 4 = 8 - 12 + 4 = 0$$

Thus, the second critical point is $(2, 0)$.

Step 4: Verify if the derivative is defined at these points Since $f(x)$ is a polynomial, its derivative $f'(x) = 3x^2 - 6x$ is defined for all real numbers x. Therefore, both critical points $(0, 4)$ and $(2, 0)$ are valid.

The function $f(x) = x^3 - 3x^2 + 4$ has two critical points:

$$(0, 4) \quad \text{and} \quad (2, 0)$$

These points are found by setting the first derivative equal to zero and solving for x, followed by substituting these values back into the original function to find the corresponding y-coordinates. Since the derivative exists at both points, they are confirmed as critical points of the function.

Further Analysis (Optional) To determine the nature of these critical points (i.e., whether they are local maxima, minima, or points of inflection), one can examine the second derivative or analyze the behavior of the first derivative around these points. However, such analysis extends beyond the scope of finding critical points and is typically covered in subsequent topics on the applications of derivatives.

3.2 Determine the intervals of concavity for $f(x) = \ln(x)$.

To determine the intervals of concavity for the function $f(x) = \ln(x)$, we analyze the second derivative of the function. Concavity describes the direction in which a function curves. A function is concave upward on intervals where its second derivative is positive and concave downward where its second derivative is negative.

First, calculate the first derivative of $f(x)$:

$$f'(x) = \frac{d}{dx} \ln(x) = \frac{1}{x}$$

Next, compute the second derivative:

$$f''(x) = \frac{d}{dx} \left(\frac{1}{x} \right) = -\frac{1}{x^2}$$

The second derivative $f''(x) = -\frac{1}{x^2}$ is always negative for all $x > 0$ since x^2 is always positive and there is a negative sign in front. Therefore, $f''(x) < 0$ for all $x > 0$.

Since the second derivative is negative on its entire domain $x > 0$, the function $f(x) = \ln(x)$ is concave downward on the interval $(0, \infty)$. There are no intervals where the function is concave upward.

Concave downward: $(0, \infty)$

There are no intervals where $f(x) = \ln(x)$ is concave upward.

3.3 Solve the optimization problem for the maximum area of a rectangle inscribed under a parabola.

Consider the problem of finding the maximum area of a rectangle that can be inscribed under the parabola defined by the equation $y = 4 - x^2$. The parabola opens downward with its vertex at the point $(0, 4)$ and intersects the x-axis at $x = -2$ and $x = 2$. The rectangle is assumed to have its base along the x-axis and its upper vertices lying on the parabola. This configuration ensures that the rectangle is symmetric with respect to the y-axis.

Let's define the coordinates of the upper right vertex of the rectangle as (x, y). Due to symmetry, the upper left vertex will be $(-x, y)$, and the base of the rectangle will span from $(-x, 0)$ to $(x, 0)$. Therefore, the width of the rectangle is $2x$, and the height is given by the value of the parabola at x, which is $y = 4 - x^2$.

The area A of the rectangle can thus be expressed as:

$$A(x) = \text{width} \times \text{height} = 2x(4 - x^2) = 8x - 2x^3$$

To find the value of x that maximizes the area $A(x)$, we need to determine the critical points of the function $A(x)$ by taking its derivative with respect to x and setting it equal to zero.

First, compute the derivative of $A(x)$:

$$A'(x) = \frac{d}{dx}(8x - 2x^3) = 8 - 6x^2$$

Set the derivative equal to zero to find the critical points:

$$8 - 6x^2 = 0 \implies 6x^2 = 8 \implies x^2 = \frac{8}{6} = \frac{4}{3} \implies x = \pm\sqrt{\frac{4}{3}} = \pm\frac{2\sqrt{3}}{3}$$

Since x represents the horizontal coordinate of a vertex in the first quadrant, we consider $x = \frac{2\sqrt{3}}{3}$.

Next, we verify that this critical point corresponds to a maximum by examining the second derivative of $A(x)$:

$$A''(x) = \frac{d}{dx}(8 - 6x^2) = -12x$$

Evaluate the second derivative at $x = \frac{2\sqrt{3}}{3}$:

$$A''\left(\frac{2\sqrt{3}}{3}\right) = -12\left(\frac{2\sqrt{3}}{3}\right) = -8\sqrt{3} < 0$$

Since the second derivative is negative, the function $A(x)$ has a local maximum at $x = \frac{2\sqrt{3}}{3}$.

Finally, substitute $x = \frac{2\sqrt{3}}{3}$ back into the expression for y to find the corresponding height:

$$y = 4 - \left(\frac{2\sqrt{3}}{3}\right)^2 = 4 - \frac{4 \times 3}{9} = 4 - \frac{12}{9} = 4 - \frac{4}{3} = \frac{12}{3} - \frac{4}{3} = \frac{8}{3}$$

Therefore, the dimensions of the rectangle that yield the maximum area are:

$$\text{Width} = 2x = 2 \times \frac{2\sqrt{3}}{3} = \frac{4\sqrt{3}}{3}$$

$$\text{Height} = \frac{8}{3}$$

The maximum area A_{max} of the rectangle is:

$$A_{\text{max}} = \text{Width} \times \text{Height} = \frac{4\sqrt{3}}{3} \times \frac{8}{3} = \frac{32\sqrt{3}}{9}$$

The rectangle with the maximum area that can be inscribed under the parabola $y = 4 - x^2$ has a width of $\frac{4\sqrt{3}}{3}$ units, a height of $\frac{8}{3}$ units, and an area of $\frac{32\sqrt{3}}{9}$ square units.

$$\boxed{\text{Maximum Area} = \frac{32\sqrt{3}}{9}}$$

3.4 Use Newton's Method to approximate the root of $x^3 - 2x + 2 = 0$.

Newton's Method, also known as the Newton-Raphson method, is an iterative numerical technique employed to find approximate solutions to equations of the form $f(x) = 0$. This method is particularly useful when analytical solutions are difficult or impossible to obtain. The fundamental principle behind Newton's Method is to start with an initial guess for the root and then iteratively refine this guess using the function and its derivative until a sufficiently accurate approximation is achieved.

Consider the cubic equation:

$$x^3 - 2x + 2 = 0$$

Our objective is to approximate one of its real roots using Newton's Method.

Step 1: Define the Function and Its Derivative

First, we define the function $f(x)$ corresponding to the given equation:

$$f(x) = x^3 - 2x + 2$$

Next, we compute the derivative of $f(x)$, denoted as $f'(x)$:

$$f'(x) = \frac{d}{dx}(x^3 - 2x + 2) = 3x^2 - 2$$

Step 2: Choose an Initial Guess

Newton's Method requires an initial guess x_0 that is reasonably close to the actual root. To select x_0, we analyze the behavior of $f(x)$ by evaluating it at various points:

$$f(-2) = (-2)^3 - 2(-2) + 2 = -8 + 4 + 2 = -2$$
$$f(-1) = (-1)^3 - 2(-1) + 2 = -1 + 2 + 2 = 3$$
$$f(0) = 0^3 - 2(0) + 2 = 0 + 0 + 2 = 2$$
$$f(1) = 1^3 - 2(1) + 2 = 1 - 2 + 2 = 1$$
$$f(2) = 2^3 - 2(2) + 2 = 8 - 4 + 2 = 6$$

From the calculations above, we observe that $f(-2) = -2$ and $f(-1) = 3$. Since the function changes sign between $x = -2$ and $x = -1$, by the Intermediate Value Theorem, there exists at least one real root in this interval. Therefore, we choose the initial guess $x_0 = -1.5$.

Step 3: Apply Newton's Iterative Formula

51

Newton's Method updates the current approximation x_n to the next approximation x_{n+1} using the formula:

$$x_{n+1} = x_n - \frac{f(x_n)}{f'(x_n)}$$

We will perform iterations until the approximate root converges to a desired level of accuracy. For demonstration purposes, we will perform three iterations.

Iteration 1:

Let $x_0 = -1.5$.

Calculate $f(x_0)$:

$$f(-1.5) = (-1.5)^3 - 2(-1.5) + 2 = -3.375 + 3 + 2 = 1.625$$

Calculate $f'(x_0)$:

$$f'(-1.5) = 3(-1.5)^2 - 2 = 3(2.25) - 2 = 6.75 - 2 = 4.75$$

Update x_1:

$$x_1 = -1.5 - \frac{1.625}{4.75} \approx -1.5 - 0.3421 = -1.8421$$

Iteration 2:

Let $x_1 \approx -1.8421$.

Calculate $f(x_1)$:

$$\begin{aligned} f(-1.8421) &= (-1.8421)^3 - 2(-1.8421) + 2 \\ &\approx -6.244 - (-3.6842) + 2 \\ &\approx -6.244 + 3.6842 + 2 = -0.5598 \end{aligned}$$

Calculate $f'(x_1)$:

$$f'(-1.8421) = 3(-1.8421)^2 - 2 \approx 3(3.393) - 2 \approx 10.179 - 2 = 8.179$$

Update x_2:

$$x_2 = -1.8421 - \frac{-0.5598}{8.179} \approx -1.8421 + 0.0685 = -1.7736$$

Iteration 3:

Let $x_2 \approx -1.7736$.

Calculate $f(x_2)$:

$$f(-1.7736) = (-1.7736)^3 - 2(-1.7736) + 2 \approx -5.568 - (-3.5472) + 2$$
$$\approx -5.568 + 3.5472 + 2$$
$$= -0.0208$$

Calculate $f'(x_2)$:

$$f'(-1.7736) = 3(-1.7736)^2 - 2 \approx 3(3.148) - 2 \approx 9.444 - 2 = 7.444$$

Update x_3:

$$x_3 = -1.7736 - \frac{-0.0208}{7.444} \approx -1.7736 + 0.0028 = -1.7708$$

Step 4: Evaluate Convergence

The change between x_2 and x_3 is $|-1.7708 - (-1.7736)| = 0.0028$. If this difference is below a predetermined tolerance level (e.g., 10^{-4}), the method can be considered to have converged to an approximate root.

For improved accuracy, additional iterations can be performed. However, in this demonstration, after three iterations, the approximate root is $x \approx -1.7708$.

Using Newton's Method with an initial guess of $x_0 = -1.5$, we have iteratively approximated a real root of the equation $x^3 - 2x + 2 = 0$ to be $x \approx -1.7708$ after three iterations. This method efficiently converges to the root, provided that the initial guess is sufficiently close and that the function satisfies the necessary conditions for convergence.

3.5 Find the local extrema of $f(x) = e^{-x^2}$.

To identify the local extrema of the function $f(x) = e^{-x^2}$, we must determine the points where the function attains local maximum or minimum values. Local extrema occur at critical points, where the first derivative of the function is either zero or undefined. In this context, since $f(x)$ is a smooth and continuous function for all real numbers x, we focus on points where the first derivative equals zero. After locating these critical points, we analyze the behavior of the function around these points to classify them as local maxima or minima.

53

First, we compute the first derivative of $f(x)$ with respect to x:

$$f'(x) = \frac{d}{dx}\left(e^{-x^2}\right).$$

Applying the chain rule for differentiation, we obtain:

$$f'(x) = e^{-x^2} \cdot (-2x) = -2xe^{-x^2}.$$

To find the critical points, we set the first derivative equal to zero and solve for x:

$$-2xe^{-x^2} = 0.$$

Since the exponential function e^{-x^2} is never zero for any real number x, the equation simplifies to:

$$-2x = 0 \quad \Rightarrow \quad x = 0.$$

Thus, the only critical point is at $x = 0$.

Next, we determine whether this critical point corresponds to a local maximum or a local minimum. To do this, we examine the second derivative of $f(x)$:

$$f''(x) = \frac{d}{dx}\left(f'(x)\right) = \frac{d}{dx}\left(-2xe^{-x^2}\right).$$

Applying the product rule to differentiate $f'(x)$, we obtain:

$$f''(x) = -2e^{-x^2} + (-2x)\cdot(-2xe^{-x^2}) = -2e^{-x^2} + 4x^2e^{-x^2} = e^{-x^2}(4x^2 - 2).$$

Evaluating the second derivative at the critical point $x = 0$:

$$f''(0) = e^{-(0)^2}(4(0)^2 - 2) = e^0(-2) = 1 \cdot (-2) = -2.$$

Since $f''(0) < 0$, the function $f(x)$ is concave downward at $x = 0$, indicating that $f(x)$ attains a local maximum at this point.

Finally, we compute the function value at $x = 0$ to determine the maximum value:

$$f(0) = e^{-(0)^2} = e^0 = 1.$$

Therefore, the function $f(x) = e^{-x^2}$ has a single local maximum at $x = 0$ with a value of 1, and there are no local minima.

3.6 Determine the inflection points of $f(x) = x^4 - 4x^3$.

An inflection point of a function is a point on the graph at which the concavity changes. To determine the inflection points of the function $f(x) = x^4 - 4x^3$, we need to analyze the second derivative of the function. The process involves finding the second derivative, identifying where it is zero or undefined, and then testing these points to confirm a change in concavity.

First, we compute the first derivative of $f(x)$:

$$f'(x) = \frac{d}{dx}(x^4 - 4x^3) = 4x^3 - 12x^2$$

Next, we find the second derivative $f''(x)$:

$$f''(x) = \frac{d}{dx}(4x^3 - 12x^2) = 12x^2 - 24x$$

To find potential inflection points, we set the second derivative equal to zero and solve for x:

$$12x^2 - 24x = 0$$

Factor out the common term $12x$:

$$12x(x - 2) = 0$$

Setting each factor equal to zero gives the critical values:

$$12x = 0 \quad \Rightarrow \quad x = 0$$

$$x - 2 = 0 \quad \Rightarrow \quad x = 2$$

These values are candidates for inflection points. To confirm that these points are indeed inflection points, we examine the sign of $f''(x)$ on intervals determined by these critical values.

Consider the intervals:

- $x < 0$

- $0 < x < 2$

- $x > 2$

55

Test point $x = -1$ in $f''(x) = 12(-1)^2 - 24(-1) = 12(1) + 24 = 36$:

$$f''(-1) = 36 > 0 \quad \text{(Concave upward)}$$

Test point $x = 1$ in $f''(x) = 12(1)^2 - 24(1) = 12 - 24 = -12$:

$$f''(1) = -12 < 0 \quad \text{(Concave downward)}$$

Test point $x = 3$ in $f''(x) = 12(3)^2 - 24(3) = 12(9) - 72 = 108 - 72 = 36$:

$$f''(3) = 36 > 0 \quad \text{(Concave upward)}$$

The sign of $f''(x)$ changes from positive to negative at $x = 0$, and from negative to positive at $x = 2$. Therefore, both $x = 0$ and $x = 2$ are inflection points.

To find the corresponding y-coordinates, substitute these x-values back into the original function $f(x)$:

For $x = 0$:
$$f(0) = (0)^4 - 4(0)^3 = 0$$

Thus, one inflection point is $(0, 0)$.

For $x = 2$:
$$f(2) = (2)^4 - 4(2)^3 = 16 - 32 = -16$$

Thus, the other inflection point is $(2, -16)$.

The function $f(x) = x^4 - 4x^3$ has two inflection points located at $(0, 0)$ and $(2, -16)$. At these points, the graph of the function changes its concavity.

3.7 Solve a related rates problem involving a ladder sliding down a wall.

Consider a scenario where a ladder is leaning against a vertical wall. The ladder is L meters long and is placed such that its base is on the ground while the top rests against the wall. As time progresses, the base of the ladder moves away from the wall, causing the top of the ladder to slide downward along the wall. This situation presents a classic related rates problem in calculus, where we aim to determine the rate at which the top of the ladder descends as the base moves away.

- **Problem Statement:**

A 10-meter-long ladder is leaning against a vertical wall. The base of the ladder is sliding away from the wall at a constant rate of 1.5 meters per second. Determine the rate at which the top of the ladder is descending along the wall at the instant when the base of the ladder is 6 meters from the wall.

- **Solution:**

To solve this problem, we will employ related rates, which involve finding the rates at which related variables change with respect to time. The relationship between the horizontal distance of the ladder's base from the wall and the vertical distance of the ladder's top from the ground is governed by the Pythagorean theorem.

Let:

- $x(t)$ be the horizontal distance from the base of the ladder to the wall at time t.

- $y(t)$ be the vertical distance from the top of the ladder to the ground at time t.

- L be the constant length of the ladder, $L = 10$ meters.

According to the Pythagorean theorem:

$$x(t)^2 + y(t)^2 = L^2$$

Differentiating both sides of the equation with respect to time t, we obtain:

$$2x(t) \cdot \frac{dx}{dt} + 2y(t) \cdot \frac{dy}{dt} = 0$$

Simplifying:

$$x(t) \cdot \frac{dx}{dt} + y(t) \cdot \frac{dy}{dt} = 0$$

We are given:

$$\frac{dx}{dt} = 1.5 \,\text{m/s}$$

$$L = 10 \,\text{m}$$

$$x(t) = 6 \,\text{m} \quad \text{(at the instant of interest)}$$

First, we need to find $y(t)$ when $x(t) = 6$ meters. Using the Pythagorean theorem:

$$6^2 + y(t)^2 = 10^2$$

57

$$36 + y(t)^2 = 100$$

$$y(t)^2 = 64$$

$$y(t) = 8\,\text{m}$$

(Note: We take the positive square root since distance cannot be negative.)

Now, substitute the known values into the differentiated equation:

$$6 \cdot 1.5 + 8 \cdot \frac{dy}{dt} = 0$$

$$9 + 8 \cdot \frac{dy}{dt} = 0$$

Solving for $\frac{dy}{dt}$:

$$8 \cdot \frac{dy}{dt} = -9$$

$$\frac{dy}{dt} = -\frac{9}{8}\,\text{m/s}$$

$$\frac{dy}{dt} = -1.125\,\text{m/s}$$

- **Interpretation:**

The negative sign indicates that $y(t)$ is decreasing over time, which means the top of the ladder is sliding downward along the wall. Therefore, at the instant when the base of the ladder is 6 meters from the wall, the top of the ladder is descending at a rate of 1.125 meters per second.

- **Conclusion:**

By applying the Pythagorean theorem and differentiating with respect to time, we established a relationship between the rates of change of the horizontal and vertical positions of the ladder. Substituting the given values allowed us to solve for the unknown rate, demonstrating how related rates can be effectively utilized to solve real-world problems involving interconnected variables.

58

3.8 Apply the Mean Value Theorem to $f(x) = x^2$ on [0,2].

The Mean Value Theorem (MVT) is a fundamental result in differential calculus that relates the average rate of change of a function over an interval to the instantaneous rate of change at some point within the interval. Specifically, the theorem states that if a function f is continuous on the closed interval $[a, b]$ and differentiable on the open interval (a, b), then there exists at least one point c in (a, b) such that

$$f'(c) = \frac{f(b) - f(a)}{b - a}.$$

This theorem essentially guarantees that for a smooth and continuous curve, there is at least one point where the tangent is parallel to the secant line connecting the endpoints of the interval.

In the context of the function $f(x) = x^2$ over the interval $[0, 2]$, we aim to apply the Mean Value Theorem to identify the specific point c where the instantaneous rate of change equals the average rate of change over the interval. To accomplish this, we must first verify that $f(x)$ satisfies the prerequisites of the theorem: continuity on $[0, 2]$ and differentiability on $(0, 2)$. Since $f(x) = x^2$ is a polynomial function, it is inherently continuous and differentiable everywhere on the real line, including the specified interval and its interior. Therefore, the conditions for applying the Mean Value Theorem are satisfied.

The next step involves calculating the average rate of change of $f(x)$ over $[0, 2]$ and then determining the value(s) of c in $(0, 2)$ where the derivative $f'(c)$ equals this average rate. The derivative $f'(x)$ represents the instantaneous rate of change of $f(x)$ at any point x.

To systematically solve the problem, we proceed with the following steps:

- Compute $f(0)$ and $f(2)$:

$$f(0) = (0)^2 = 0,$$

$$f(2) = (2)^2 = 4.$$

- Determine the average rate of change over $[0, 2]$:

$$\frac{f(2) - f(0)}{2 - 0} = \frac{4 - 0}{2} = \frac{4}{2} = 2.$$

- Find the derivative $f'(x)$:

$$f'(x) = \frac{d}{dx}x^2 = 2x.$$

59

- Set $f'(c)$ equal to the average rate of change and solve for c:

$$f'(c) = 2c = 2.$$

$$2c = 2 \implies c = 1.$$

- Verify that $c = 1$ lies within the open interval $(0, 2)$: Since 1 is greater than 0 and less than 2, it satisfies the condition $c \in (0, 2)$.

Thus, $c = 1$ is the specific point in the interval $(0, 2)$ where the instantaneous rate of change of $f(x) = x^2$ equals the average rate of change over the interval, as guaranteed by the Mean Value Theorem.

Applying the Mean Value Theorem to $f(x) = x^2$ on $[0, 2]$ confirms that there exists a point $c = 1$ within the interval where the derivative $f'(c) = 2$ matches the average rate of change of 2 over the interval. This outcome exemplifies the theorem's assurance of the existence of such a point under the given conditions.

3.9 Analyze the motion of a particle with position $s(t) = t^3 - 6t^2 + 9t$.

The motion of a particle along a straight line can be comprehensively understood by analyzing its position function with respect to time. Given the position function $s(t) = t^3 - 6t^2 + 9t$, we aim to determine the particle's velocity, acceleration, critical points, intervals of increasing or decreasing velocity, and the overall behavior of the motion over time. This analysis involves calculating the first and second derivatives of the position function, examining their properties, and interpreting the results in the context of kinematics.

To begin, the position function $s(t)$ describes the location of the particle at any time t. The first derivative of $s(t)$ with respect to time t gives the velocity function $v(t)$, which represents the rate of change of position, indicating how fast and in which direction the particle is moving. The second derivative, $a(t)$, obtained by differentiating the velocity function, represents the acceleration of the particle, describing how the velocity is changing over time.

$$v(t) = \frac{ds}{dt} = \frac{d}{dt}\left(t^3 - 6t^2 + 9t\right) = 3t^2 - 12t + 9$$

$$a(t) = \frac{dv}{dt} = \frac{d}{dt}\left(3t^2 - 12t + 9\right) = 6t - 12$$

To identify the critical points where the velocity is zero, we set $v(t) = 0$:

$$3t^2 - 12t + 9 = 0$$

Dividing both sides by 3 simplifies the equation:

$$t^2 - 4t + 3 = 0$$

Factoring the quadratic equation:

$$(t - 1)(t - 3) = 0$$

Thus, the critical points occur at $t = 1$ and $t = 3$.

Next, we examine the intervals determined by these critical points to identify where the velocity is positive (indicating motion in the positive direction) or negative (indicating motion in the negative direction). We consider three intervals: $t < 1$, $1 < t < 3$, and $t > 3$.

- For $t < 1$: Choose $t = 0$:

$$v(0) = 3(0)^2 - 12(0) + 9 = 9 > 0$$

  The velocity is positive, indicating the particle is moving in the positive direction.

- For $1 < t < 3$: Choose $t = 2$:

$$v(2) = 3(2)^2 - 12(2) + 9 = 12 - 24 + 9 = -3 < 0$$

  The velocity is negative, indicating the particle is moving in the negative direction.

- For $t > 3$: Choose $t = 4$:

$$v(4) = 3(4)^2 - 12(4) + 9 = 48 - 48 + 9 = 9 > 0$$

  The velocity is positive, indicating the particle is moving again in the positive direction.

To determine the acceleration and understand how it affects the velocity, we analyze the acceleration function $a(t) = 6t - 12$. Setting $a(t) = 0$ helps identify when the acceleration changes sign:

$$6t - 12 = 0 \implies t = 2$$

This critical time divides the acceleration behavior into two intervals: $t < 2$ and $t > 2$.

- For $t < 2$: Choose $t = 1$:
$$a(1) = 6(1) - 12 = -6 < 0$$

  The acceleration is negative, indicating that the velocity is decreasing over time.

- For $t > 2$: Choose $t = 3$:
$$a(3) = 6(3) - 12 = 6 > 0$$

  The acceleration is positive, indicating that the velocity is increasing over time.

Combining the information about velocity and acceleration, we can describe the particle's motion:

- From $t = 0$ to $t = 1$: The velocity is positive, and acceleration is negative. This means the particle is moving forward but slowing down as it approaches $t = 1$.

- At $t = 1$: The velocity becomes zero, indicating a momentary stop before changing direction.

- From $t = 1$ to $t = 3$: The velocity is negative, and acceleration is negative until $t = 2$, then positive after $t = 2$. Initially, the particle moves backward, slowing down until $t = 2$, where the acceleration becomes positive, causing the velocity to increase in the negative direction (i.e., the particle accelerates backward), until $t = 3$.

- At $t = 3$: The velocity returns to zero, indicating another momentary stop.

- From $t = 3$ onward: The velocity is positive, and acceleration remains positive. The particle moves forward, speeding up as both velocity and acceleration are positive.

In summary, the position function $s(t) = t^3 - 6t^2 + 9t$ describes a particle that undergoes changes in direction and speed over time. By analyzing the first and second derivatives, we identify critical points, intervals of motion direction, and how acceleration influences the velocity, providing a comprehensive understanding of the particle's motion dynamics.

3.10   Find the maximum profit given a cost and revenue function.

In this problem, we aim to determine the level of production that maximizes profit for a particular product. Profit is defined as the difference between total revenue and total cost. Specifically, given the revenue function $R(x)$ and the cost function $C(x)$, where x represents the number of units produced and sold, the profit function $P(x)$ is expressed as:

$$P(x) = R(x) - C(x)$$

Our objective is to find the value of x that maximizes $P(x)$. This involves analyzing the profit function to identify its critical points and determining which of these points correspond to maximum profit.

Problem Statement: Consider a company that manufactures and sells x units of a product. The revenue generated from selling x units is given by the function:

$$R(x) = -5x^2 + 150x$$

The total cost of producing x units is given by:

$$C(x) = 20x + 500$$

Determine the number of units x that the company should produce and sell to achieve maximum profit. Additionally, calculate the maximum profit.

Solution:

To find the production level x that maximizes profit, we will perform the following steps:

- Define the Profit Function:

$$P(x) = R(x) - C(x)$$

- Differentiate the Profit Function with respect to x:

$$P'(x) = \frac{dP}{dx}$$

- Find Critical Points by Setting $P'(x) = 0$:

$$P'(x) = 0$$

63

- Determine the Nature of Critical Points (Maximum or Minimum) Using the Second Derivative Test:

$$P''(x) = \frac{d^2 P}{dx^2}$$

- Compute the Maximum Profit by Substituting the Critical Point into $P(x)$.

Let's carry out each step in detail.

1. Define the Profit Function

Given the revenue and cost functions:

$$R(x) = -5x^2 + 150x$$

$$C(x) = 20x + 500$$

The profit function is:

$$P(x) = R(x) - C(x) = (-5x^2 + 150x) - (20x + 500) = -5x^2 + 130x - 500$$

2. Differentiate the Profit Function with respect to x

To find the critical points, we first compute the first derivative of $P(x)$:

$$P'(x) = \frac{d}{dx}(-5x^2 + 130x - 500) = -10x + 130$$

3. Find Critical Points by Setting $P'(x) = 0$

Setting the first derivative equal to zero to find critical points:

$$-10x + 130 = 0$$

Solving for x:

$$-10x = -130 \implies x = \frac{130}{10} = 13$$

Thus, $x = 13$ is the critical point.

4. Determine the Nature of Critical Points Using the Second Derivative Test

To ascertain whether $x = 13$ yields a maximum or minimum profit, we examine the second derivative of $P(x)$:

$$P''(x) = \frac{d}{dx}(-10x + 130) = -10$$

Since $P''(x) = -10 < 0$, the profit function is concave downward at $x = 13$. Therefore, $x = 13$ corresponds to a local maximum.

5. Compute the Maximum Profit

Substituting $x = 13$ into the profit function $P(x)$:

$$P(13) = -5(13)^2 + 130(13) - 500$$

Calculating each term:

$$13^2 = 169$$

$$-5 \times 169 = -845$$

$$130 \times 13 = 1690$$

Now, summing the terms:

$$P(13) = -845 + 1690 - 500 = 345$$

Thus, the maximum profit is $345.

The company should produce and sell 13 units of the product to achieve the maximum profit of $345. This conclusion is derived by formulating the profit function, finding its critical point through differentiation, confirming the nature of the critical point using the second derivative, and finally calculating the profit at that critical point.

Chapter 4

Integration Techniques

This chapter delves into various integration techniques to equip readers with the tools necessary for solving complex integrals. It includes problems focused on methods such as substitution, integration by parts, partial fraction decomposition, and trigonometric integrals. Each problem is designed to enhance computational skills and deepen conceptual understanding. The solutions provided offer step-by-step guidance to elucidate the intricacies of different techniques, fostering a robust grasp of integral calculus that will be essential for tackling more advanced topics.

4.1 Evaluate $\int \frac{1}{x \ln(x)} dx$.

The integral $\int \frac{1}{x \ln(x)} dx$ presents a challenge that requires the application of substitution, a fundamental integration technique. The integrand consists of a rational function where the denominator includes a logarithmic expression. Recognizing the relationship between the derivative of the logarithmic function and the components of the integrand is crucial for simplifying and evaluating the integral.

To approach this problem, we observe that the denominator $\ln(x)$ suggests a natural substitution involving the logarithm of x. Specifically, if we let $u = \ln(x)$, then the differential du can be expressed in terms of dx, facilitating the transformation of the integral into a more manageable form.

Starting with the substitution:

$$u = \ln(x)$$

Differentiating both sides with respect to x yields:

$$\frac{du}{dx} = \frac{1}{x} \quad \Rightarrow \quad du = \frac{1}{x}dx$$

This substitution directly aligns with the $\frac{1}{x}dx$ term present in the integrand. Substituting u and du into the original integral transforms it as follows:

$$\int \frac{1}{x\ln(x)}dx = \int \frac{1}{u}du$$

The transformed integral $\int \frac{1}{u}du$ is a standard logarithmic integral. Evaluating this integral, we obtain:

$$\int \frac{1}{u}du = \ln|u| + C$$

Here, C represents the constant of integration, accounting for the family of antiderivatives.

Substituting back the original substitution $u = \ln(x)$, the expression becomes:

$$\ln|\ln(x)| + C$$

Since $x > 0$ in the domain of $\ln(x)$, the absolute value can be omitted, simplifying the final expression to:

$$\ln(\ln(x)) + C$$

Therefore, the evaluated integral is:

$$\int \frac{1}{x\ln(x)}dx = \ln(\ln(x)) + C$$

This solution demonstrates the effective use of substitution to simplify and evaluate an integral involving a logarithmic function in the denominator.

4.2 Compute $\int e^{2x}\sin(3x)\,dx$.

The integral $\int e^{2x}\sin(3x)\,dx$ presents a challenge that involves the interplay between exponential and trigonometric functions. To evaluate this integral, one must employ the technique of integration by parts, which is particularly useful when dealing with products of functions whose individual integrals and derivatives are known. Integration by parts is based on the product rule for differentiation and is formally stated as:

$$\int u\,dv = uv - \int v\,du$$

In this context, choosing appropriate functions for u and dv is crucial for simplifying the integral into a solvable form. Given the functions e^{2x} and $\sin(3x)$, a strategic selection will allow the integral to be expressed in terms of itself, facilitating the solution through algebraic manipulation.

Let us proceed with the step-by-step evaluation of the integral $\int e^{2x} \sin(3x)\,dx$:

First, we apply integration by parts. We choose:

$$\text{Let } u = e^{2x} \quad \Rightarrow \quad du = 2e^{2x}\,dx$$

$$dv = \sin(3x)\,dx \quad \Rightarrow \quad v = -\frac{1}{3}\cos(3x)$$

Substituting into the integration by parts formula:

$$\int e^{2x} \sin(3x)\,dx = uv - \int v\,du = e^{2x}\left(-\frac{1}{3}\cos(3x)\right) - \int \left(-\frac{1}{3}\cos(3x)\right)\cdot 2e^{2x}\,dx$$

Simplifying:

$$= -\frac{1}{3}e^{2x}\cos(3x) + \frac{2}{3}\int e^{2x}\cos(3x)\,dx$$

We now denote the original integral as I:

$$I = \int e^{2x} \sin(3x)\,dx$$

From the previous step, we have:

$$I = -\frac{1}{3}e^{2x}\cos(3x) + \frac{2}{3}\int e^{2x}\cos(3x)\,dx$$

Next, we must evaluate the integral $\int e^{2x}\cos(3x)\,dx$. We apply integration by parts again, selecting:

$$\text{Let } u = e^{2x} \quad \Rightarrow \quad du = 2e^{2x}\,dx$$

$$dv = \cos(3x)\,dx \quad \Rightarrow \quad v = \frac{1}{3}\sin(3x)$$

Substituting into the integration by parts formula:

$$\int e^{2x} \cos(3x)\,dx = uv - \int v\,du = e^{2x}\left(\frac{1}{3}\sin(3x)\right) - \int \left(\frac{1}{3}\sin(3x)\right)\cdot 2e^{2x}\,dx$$

69

Simplifying:

$$= \frac{1}{3}e^{2x}\sin(3x) - \frac{2}{3}\int e^{2x}\sin(3x)\ dx$$

Notice that the integral $\int e^{2x}\sin(3x)\ dx$ reappears, which we've denoted as I. Substituting back:

$$\int e^{2x}\cos(3x)\ dx = \frac{1}{3}e^{2x}\sin(3x) - \frac{2}{3}I$$

Returning to our earlier expression for I:

$$I = -\frac{1}{3}e^{2x}\cos(3x) + \frac{2}{3}\left(\frac{1}{3}e^{2x}\sin(3x) - \frac{2}{3}I\right)$$

Expanding the terms:

$$I = -\frac{1}{3}e^{2x}\cos(3x) + \frac{2}{9}e^{2x}\sin(3x) - \frac{4}{9}I$$

To solve for I, we collect like terms:

$$I + \frac{4}{9}I = -\frac{1}{3}e^{2x}\cos(3x) + \frac{2}{9}e^{2x}\sin(3x)$$

$$\frac{13}{9}I = -\frac{1}{3}e^{2x}\cos(3x) + \frac{2}{9}e^{2x}\sin(3x)$$

Multiplying both sides by $\frac{9}{13}$ to isolate I:

$$I = \frac{9}{13}\left(-\frac{1}{3}e^{2x}\cos(3x) + \frac{2}{9}e^{2x}\sin(3x)\right)$$

$$I = -\frac{3}{13}e^{2x}\cos(3x) + \frac{2}{13}e^{2x}\sin(3x) + C$$

Where C represents the constant of integration. Therefore, the evaluated integral is:

$$\int e^{2x}\sin(3x)\ dx = -\frac{3}{13}e^{2x}\cos(3x) + \frac{2}{13}e^{2x}\sin(3x) + C$$

This solution exemplifies the application of integration by parts, particularly when dealing with products of exponential and trigonometric functions. By strategically choosing u and dv, and recognizing the reappearance of the original integral within the process, we can solve for I algebraically, resulting in a complete and precise evaluation of the integral.

4.3 Evaluate $\int \frac{dx}{x^2+a^2}$.

The integral $\int \frac{dx}{x^2+a^2}$ represents a fundamental form in integral calculus, commonly encountered in various applications within mathematics and the physical sciences. This integral is significant due to its appearance in the process of solving differential equations, evaluating areas under curves, and modeling phenomena such as oscillations and waveforms. Understanding the method to evaluate this integral is crucial for mastering integration techniques, particularly those involving rational functions and trigonometric identities.

To evaluate $\int \frac{dx}{x^2+a^2}$, we recognize that the integrand is a rational function where the denominator is a sum of squares. This form suggests the application of a standard integral formula involving the inverse tangent function. The integral can be approached through a straightforward substitution that aligns the integrand with the derivative of the inverse tangent function, thereby facilitating its evaluation.

$$\int \frac{dx}{x^2 + a^2}$$

Let's proceed with a meticulous step-by-step solution:

- **Step 1: Identify the Integral Form** The integral $\int \frac{dx}{x^2+a^2}$ is of the form $\int \frac{dx}{x^2+a^2}$, where a is a constant. This corresponds to a standard integral formula involving the inverse tangent function.

- **Step 2: Recall the Standard Integral Formula** The standard integral formula is:
$$\int \frac{du}{u^2 + c^2} = \frac{1}{c} \tan^{-1} \left(\frac{u}{c} \right) + C$$
where C is the constant of integration.

- **Step 3: Compare the Given Integral with the Standard Form** In the given integral, $u = x$ and $c = a$. Therefore, we can directly apply the standard formula without the need for substitution.

- **Step 4: Apply the Standard Integral Formula** Applying the formula:
$$\int \frac{dx}{x^2 + a^2} = \frac{1}{a} \tan^{-1} \left(\frac{x}{a} \right) + C$$

- **Step 5: Simplify the Expression** The expression is already simplified, and the integral has been evaluated in terms of the inverse tangent function.

71

Final Solution:

$$\int \frac{dx}{x^2 + a^2} = \frac{1}{a} \tan^{-1}\left(\frac{x}{a}\right) + C$$

Verification Through Differentiation: To verify the correctness of the solution, we differentiate the result with respect to x:

$$\frac{d}{dx}\left(\frac{1}{a} \tan^{-1}\left(\frac{x}{a}\right) + C\right) = \frac{1}{a} \cdot \frac{1}{1 + \left(\frac{x}{a}\right)^2} \cdot \frac{1}{a} = \frac{1}{x^2 + a^2}$$

This confirms that the derivative of the evaluated integral yields the original integrand, thereby validating the solution.

Conclusion: The integral $\int \frac{dx}{x^2 + a^2}$ is evaluated using the standard inverse tangent formula, resulting in $\frac{1}{a} \tan^{-1}\left(\frac{x}{a}\right) + C$. This method exemplifies the application of fundamental integral formulas to solve rational functions involving sums of squares, a technique that is widely applicable in various branches of mathematics and physics.

4.4 Compute $\int \ln(x)\, dx$.

The integral of the natural logarithm function, $\int \ln(x)\, dx$, is a fundamental problem in calculus that serves to illustrate the application of integration techniques, specifically integration by parts. The natural logarithm function, $\ln(x)$, arises frequently in various fields such as mathematics, physics, and engineering, making its integral a crucial component in solving more complex problems. Understanding the process of integrating $\ln(x)$ not only enhances computational skills but also deepens the conceptual grasp of integral calculus.

To evaluate $\int \ln(x)\, dx$, we employ the method of integration by parts, which is based on the product rule for differentiation. Integration by parts is particularly useful when the integrand is the product of two functions whose individual integrals are known or can be easily determined. The formula for integration by parts is given by:

$$\int u\, dv = uv - \int v\, du$$

In this context, we need to identify appropriate choices for u and dv such that their differentiation and integration, respectively, simplify the integral.

- **Step 1: Choose u and dv**

Let us assign:

$$u = \ln(x)$$
$$dv = dx$$

- **Step 2: Differentiate u and Integrate dv**

  Differentiating u with respect to x, we obtain:

  $$du = \frac{d}{dx}\ln(x)\,dx = \frac{1}{x}\,dx$$

  Integrating dv with respect to x, we find:

  $$v = \int dv = \int dx = x$$

- **Step 3: Apply the Integration by Parts Formula**

  Substituting u, du, v, and dv into the integration by parts formula:

  $$\int \ln(x)\,dx = uv - \int v\,du$$

  $$= \ln(x) \cdot x - \int x \cdot \frac{1}{x}\,dx$$

- **Step 4: Simplify the Integral**

  Simplify the integrand in the remaining integral:

  $$\int x \cdot \frac{1}{x}\,dx = \int 1\,dx$$

- **Step 5: Integrate the Simplified Expression**

  Integrate 1 with respect to x:

  $$\int 1\,dx = x + C$$

  where C is the constant of integration.

- **Step 6: Combine the Results**

  Substituting back into the expression obtained from integration by parts:

  $$\int \ln(x)\,dx = x\ln(x) - (x + C)$$

73

- **Step 7: Simplify the Final Expression**

  Combine like terms to express the integral in its simplest form:

  $$\int \ln(x)\,dx = x\ln(x) - x + C$$

Conclusion

The integral of the natural logarithm function with respect to x is given by:

$$\int \ln(x)\,dx = x\ln(x) - x + C$$

where C is the constant of integration. This result demonstrates the effectiveness of the integration by parts technique in evaluating integrals involving logarithmic functions.

4.5 Evaluate $\int \tan(x)\,dx$.

The integral $\int \tan(x)\,dx$ requires the application of fundamental integration techniques to determine its antiderivative. The tangent function, $\tan(x)$, can be expressed in terms of sine and cosine functions, which facilitates the integration process. Recognizing the relationship between $\tan(x)$ and $\cos(x)$ is crucial for selecting an appropriate substitution that simplifies the integral into a more manageable form.

To evaluate $\int \tan(x)\,dx$, we begin by expressing the tangent function as the ratio of sine and cosine:

$$\tan(x) = \frac{\sin(x)}{\cos(x)}$$

Substituting this into the integral, we have:

$$\int \tan(x)\,dx = \int \frac{\sin(x)}{\cos(x)}\,dx$$

This formulation suggests the use of a substitution method, specifically choosing $u = \cos(x)$, since the derivative of $\cos(x)$ is directly related to $\sin(x)$, which appears in the numerator of the integrand. Let us proceed with this substitution:

74

$$u = \cos(x)$$
$$\Rightarrow \quad \frac{du}{dx} = -\sin(x)$$
$$\Rightarrow \quad du = -\sin(x)\,dx$$
$$\sin(x)\,dx = -du$$

Substituting u and du back into the integral, we replace $\cos(x)$ with u and $\sin(x)\,dx$ with $-du$:

$$\int \frac{\sin(x)}{\cos(x)}\,dx = \int \frac{-du}{u} = -\int \frac{1}{u}\,du$$

The integral $\int \frac{1}{u}\,du$ is a standard logarithmic integral, whose antiderivative is the natural logarithm of the absolute value of u:

$$-\int \frac{1}{u}\,du = -\ln|u| + C$$

Substituting back the original expression for u, which is $\cos(x)$, we obtain:

$$-\ln|\cos(x)| + C$$

Recognizing that $\ln|\cos(x)|$ is equivalent to $-\ln|\sec(x)|$ due to the reciprocal identity $\sec(x) = \frac{1}{\cos(x)}$, we can simplify the expression further:

$$-\ln|\cos(x)| + C = \ln|\sec(x)| + C$$

Therefore, the integral of $\tan(x)$ with respect to x is:

$$\int \tan(x)\,dx = -\ln|\cos(x)| + C = \ln|\sec(x)| + C$$

where C is the constant of integration. This result provides the general antiderivative of the tangent function, encapsulating all possible antiderivatives differing by a constant.

$$\boxed{\int \tan(x)\,dx = -\ln|\cos(x)| + C}$$

This solution demonstrates the application of substitution in integrating trigonometric functions, particularly tangent, by leveraging the relationship between sine and cosine to simplify the integral into a logarithmic form. The method employed ensures a clear and systematic approach, enhancing the learner's understanding of integration techniques involving trigonometric identities and substitution.

4.6 Compute $\int xe^{x^2}\, dx$.

The integral $\int xe^{x^2}\, dx$ presents a challenge due to the composition of functions involving an exponential function with a quadratic argument multiplied by a linear term. To evaluate this integral, one must employ an appropriate integration technique that simplifies the expression into a form that can be readily integrated. The method of substitution, also known as u-substitution, is particularly effective in this context. By identifying a suitable substitution that reduces the integral to a basic form, the computation becomes more manageable. This approach leverages the relationship between the derivative of the exponent and the algebraic factor present in the integrand, facilitating the integration process.

To begin, consider the integral $\int xe^{x^2}\, dx$. Notice that the exponent x^2 is a function whose derivative with respect to x is $2x$, which is closely related to the algebraic term x present in the integrand. This observation suggests that a substitution involving the exponent could simplify the integral. Define the substitution $u = x^2$, which implies that $\frac{du}{dx} = 2x$ or equivalently, $du = 2x\, dx$. Solving for $x\, dx$, we find $x\, dx = \frac{1}{2} du$.

Substituting these expressions into the original integral transforms it as follows:

$$\int xe^{x^2}\, dx = \int e^u \cdot \frac{1}{2}\, du = \frac{1}{2} \int e^u\, du$$

The integral of e^u with respect to u is straightforward:

$$\frac{1}{2} \int e^u\, du = \frac{1}{2}e^u + C$$

Finally, reverting the substitution by replacing u with x^2 yields the solution in terms of the original variable:

$$\frac{1}{2}e^{x^2} + C$$

where C represents the constant of integration. This result demonstrates that the integral of xe^{x^2} with respect to x is $\frac{1}{2}e^{x^2} + C$.

$$\int xe^{x^2}\, dx = \frac{1}{2}e^{x^2} + C$$

This solution highlights the effectiveness of substitution in integrating products of algebraic and exponential functions, especially when the exponent's derivative is a scalar multiple of an existing factor in the integrand. Understanding and identifying such relationships are crucial for simplifying and solving complex integrals encountered in calculus.

4.7 Evaluate $\int \frac{dx}{\sqrt{a^2-x^2}}$.

The integral $\int \frac{dx}{\sqrt{a^2-x^2}}$ is a standard form encountered in integral calculus, particularly within the context of inverse trigonometric functions. This integral arises frequently in various applications, including geometry, physics, and engineering, where it may represent the length of an arc or the solution to certain differential equations. The integrand $\frac{1}{\sqrt{a^2-x^2}}$ suggests a relationship to the Pythagorean identity, which is fundamental in trigonometric substitution methods.

To evaluate this integral, we can employ a direct substitution method by recognizing it as the derivative of an inverse sine function. Specifically, recall that the derivative of $\arcsin\left(\frac{x}{a}\right)$ with respect to x is $\frac{1}{\sqrt{a^2-x^2}}$. This relationship indicates that the antiderivative of $\frac{1}{\sqrt{a^2-x^2}}$ is $\arcsin\left(\frac{x}{a}\right)$, plus a constant of integration.

Alternatively, one can approach the integral using trigonometric substitution to reinforce the connection with inverse trigonometric functions. Let us perform a substitution where $x = a\sin\theta$. Consequently, $dx = a\cos\theta\, d\theta$, and the denominator becomes $\sqrt{a^2 - a^2\sin^2\theta} = a\cos\theta$. Substituting these into the integral transforms it into:

$$\int \frac{a\cos\theta\, d\theta}{a\cos\theta} = \int d\theta = \theta + C$$

77

Since $x = a\sin\theta$, we solve for θ to obtain $\theta = \arcsin\left(\frac{x}{a}\right)$. Substituting back, the integral simplifies to:

$$\arcsin\left(\frac{x}{a}\right) + C$$

Thus, the original integral $\int \frac{dx}{\sqrt{a^2-x^2}}$ evaluates to $\arcsin\left(\frac{x}{a}\right) + C$, where C represents the constant of integration. This result aligns with the fundamental antiderivative of the inverse sine function, confirming the correctness of the evaluation through both direct recognition and substitution methods.

$$\int \frac{dx}{\sqrt{a^2-x^2}} = \arcsin\left(\frac{x}{a}\right) + C$$

This comprehensive approach not only provides the solution but also elucidates the underlying principles connecting trigonometric identities and inverse functions in the context of integration techniques.

4.8 Compute $\int \frac{dx}{x^2\sqrt{x^2-1}}$.

The integral $\int \frac{dx}{x^2\sqrt{x^2-1}}$ presents a challenge due to the presence of both a rational function and a square root of a quadratic expression. To evaluate this integral, it is necessary to employ an appropriate substitution that simplifies the integrand into a form that is more manageable. A strategic choice of substitution can transform the integral into one that is readily solvable using standard calculus techniques.

We begin by observing the structure of the integrand. The expression under the square root, x^2-1, suggests the use of a trigonometric substitution, particularly because it resembles the form $\sec^2\theta - 1 = \tan^2\theta$. This resemblance indicates that setting $x = \sec\theta$ could simplify the square root term.

- **Step 1: Trigonometric Substitution**

  Let us make the substitution:

  $$x = \sec\theta$$

  Differentiating both sides with respect to θ:

  $$dx = \sec\theta\tan\theta\,d\theta$$

Next, we express $\sqrt{x^2-1}$ in terms of θ:

$$\sqrt{x^2-1} = \sqrt{\sec^2\theta - 1} = \sqrt{\tan^2\theta} = \tan\theta$$

Substituting x and dx into the original integral:

$$\int \frac{dx}{x^2\sqrt{x^2-1}} = \int \frac{\sec\theta\tan\theta\, d\theta}{\sec^2\theta\cdot\tan\theta} = \int \frac{\sec\theta\tan\theta}{\sec^2\theta\tan\theta}\, d\theta$$

- **Step 2: Simplification of the Integrand**

Notice that $\tan\theta$ appears in both the numerator and the denominator, and $\sec^2\theta$ in the denominator cancels with $\sec\theta$ in the numerator:

$$\int \frac{\sec\theta\tan\theta}{\sec^2\theta\tan\theta}\, d\theta = \int \frac{1}{\sec\theta}\, d\theta = \int \cos\theta\, d\theta$$

This simplification reduces the integral to a basic trigonometric integral:

$$\int \cos\theta\, d\theta = \sin\theta + C$$

where C is the constant of integration.

- **Step 3: Back-Substitution to Original Variable**

To express the solution in terms of the original variable x, we revert to the substitution $x = \sec\theta$. From this substitution, we have:

$$\theta = \sec^{-1} x$$

and from the trigonometric identity,

$$\sin\theta = \frac{\sqrt{x^2-1}}{x}$$

Thus, substituting back:

$$\sin\theta = \frac{\sqrt{x^2-1}}{x}$$

Therefore, the integral in terms of x is:

$$\sin\theta + C = \frac{\sqrt{x^2-1}}{x} + C$$

where C remains the constant of integration.

79

Final Answer:

$$\int \frac{dx}{x^2\sqrt{x^2-1}} = \frac{\sqrt{x^2-1}}{x} + C$$

This solution demonstrates the effectiveness of trigonometric substitution in evaluating integrals involving square roots of quadratic expressions. By carefully selecting an appropriate substitution, the integral is transformed into a form that is straightforward to integrate, ultimately leading to a solution expressed in terms of the original variable.

4.9 Evaluate the improper integral $\int_1^\infty \frac{dx}{x^p}$ for $p > 1$.

The integral $\int_1^\infty \frac{dx}{x^p}$ is classified as an improper integral due to its infinite upper limit of integration. In evaluating such integrals, it is essential to determine whether the integral converges to a finite value or diverges to infinity. The behavior of the integrand $\frac{1}{x^p}$ as x approaches infinity plays a crucial role in this determination.

For $p > 1$, the function $\frac{1}{x^p}$ decreases rapidly enough as x increases, suggesting that the area under the curve from $x = 1$ to $x = \infty$ may converge. Conversely, if $p \leq 1$, the decrease is insufficient to guarantee convergence, and the integral may diverge.

To evaluate the integral, we express it as a limit:

$$\int_1^\infty \frac{dx}{x^p} = \lim_{b \to \infty} \int_1^b \frac{dx}{x^p}$$

Applying the power rule for integration, which states that $\int x^n \, dx = \frac{x^{n+1}}{n+1} + C$ for $n \neq -1$, we integrate $\frac{1}{x^p} = x^{-p}$:

$$\int x^{-p} \, dx = \frac{x^{-p+1}}{-p+1} + C = \frac{x^{1-p}}{1-p} + C$$

Evaluating the definite integral from 1 to b:

$$\int_1^b \frac{dx}{x^p} = \left[\frac{x^{1-p}}{1-p} \right]_1^b = \frac{b^{1-p}}{1-p} - \frac{1^{1-p}}{1-p} = \frac{b^{1-p}-1}{1-p}$$

Simplifying the expression:

80

$$\frac{b^{1-p}-1}{1-p} = \frac{1-b^{1-p}}{p-1}$$

Taking the limit as b approaches infinity:

$$\lim_{b\to\infty} \frac{1-b^{1-p}}{p-1}$$

Since $p > 1$, the exponent $1-p$ is negative, and $b^{1-p} = \frac{1}{b^{p-1}}$. As b approaches infinity, $\frac{1}{b^{p-1}}$ approaches zero. Therefore, the limit simplifies to:

$$\lim_{b\to\infty} \frac{1-0}{p-1} = \frac{1}{p-1}$$

Thus, the improper integral converges to $\frac{1}{p-1}$ when $p > 1$:

$$\int_1^\infty \frac{dx}{x^p} = \frac{1}{p-1}$$

This result demonstrates that for values of p greater than 1, the area under the curve $\frac{1}{x^p}$ from $x = 1$ to $x = \infty$ is finite and equal to $\frac{1}{p-1}$. Conversely, if $p \leq 1$, the integral does not converge, highlighting the critical role of the exponent p in determining the behavior of the improper integral.

4.10 Compute $\int \frac{x^2}{(x^3+1)^2}\,dx.$

The integral $\int \frac{x^2}{(x^3+1)^2}\,dx$ presents a rational function where the polynomial in the numerator is related to the derivative of the polynomial in the denominator. To evaluate this integral, we can employ the method of substitution, which is particularly effective when the integrand contains a function and its derivative.

Let us denote:

$$u = x^3 + 1$$

Differentiating both sides with respect to x gives:

$$\frac{du}{dx} = 3x^2$$

This implies:

$$du = 3x^2\,dx \quad \Rightarrow \quad x^2\,dx = \frac{du}{3}$$

Substituting u and du into the original integral:

$$\int \frac{x^2}{(x^3+1)^2}\,dx = \int \frac{1}{u^2}\cdot\frac{du}{3} = \frac{1}{3}\int u^{-2}\,du$$

Integrating u^{-2} with respect to u:

$$\frac{1}{3}\int u^{-2}\,du = \frac{1}{3}\left(\frac{u^{-1}}{-1}\right)+C = -\frac{1}{3}u^{-1}+C = -\frac{1}{3u}+C$$

Finally, substituting back $u = x^3 + 1$:

$$-\frac{1}{3u}+C = -\frac{1}{3(x^3+1)}+C$$

Therefore, the evaluated integral is:

$$\int \frac{x^2}{(x^3+1)^2}\,dx = -\frac{1}{3(x^3+1)}+C$$

where C is the constant of integration.

Chapter 5

Applications of Integrals

This chapter focuses on the diverse applications of integrals, encompassing a range of problems that illustrate their practical importance. Topics include calculating areas under curves, volumes of solids of revolution, work done by a force, and applications in physics and engineering. Through these problems, readers will gain insight into how integrals are used to model and solve real-world situations. Each problem is complemented by thorough solutions, aimed at reinforcing comprehension and highlighting the relevance of integrals in various contexts.

5.1 Find the area between $y = x^2$ and $y = x + 2$.

To determine the area enclosed between the curves $y = x^2$ and $y = x + 2$, we must first identify the points of intersection between the two functions. This requires solving the equation $x^2 = x + 2$. Rearranging terms leads to the quadratic equation $x^2 - x - 2 = 0$. Applying the quadratic formula:

$$x = \frac{1 \pm \sqrt{1 + 8}}{2} = \frac{1 \pm 3}{2}$$

Thus, the solutions are $x = 2$ and $x = -1$. These intersection points indicate the interval over which the area between the curves will be calculated.

Next, we must determine which function lies above the other within the interval $[-1, 2]$. By selecting a test point within this interval, such as $x = 0$:

$$y = (0)^2 = 0 \quad \text{and} \quad y = 0 + 2 = 2$$

Since $y = x + 2$ yields a higher value than $y = x^2$ at $x = 0$, it follows that $y = x + 2$ is the upper function and $y = x^2$ is the lower function over the interval $[-1, 2]$.

The area A between the two curves can be expressed as the definite integral of the difference between the upper and lower functions over the interval from $x = -1$ to $x = 2$:

$$A = \int_{-1}^{2} \left((x + 2) - x^2 \right) dx$$

Simplifying the integrand:

$$A = \int_{-1}^{2} (-x^2 + x + 2)\, dx$$

We proceed by integrating each term separately:

$$\int (-x^2)\, dx = -\frac{x^3}{3}, \quad \int x\, dx = \frac{x^2}{2}, \quad \int 2\, dx = 2x$$

Combining these results, the antiderivative $F(x)$ is:

$$F(x) = -\frac{x^3}{3} + \frac{x^2}{2} + 2x$$

We then evaluate $F(x)$ at the upper limit $x = 2$ and the lower limit $x = -1$, and subtract accordingly:

$$A = \left[-\frac{(2)^3}{3} + \frac{(2)^2}{2} + 2(2) \right] - \left[-\frac{(-1)^3}{3} + \frac{(-1)^2}{2} + 2(-1) \right]$$

Calculating each term:

For $x = 2$:

$$-\frac{8}{3} + \frac{4}{2} + 4 = -\frac{8}{3} + 2 + 4 = -\frac{8}{3} + 6 = \frac{10}{3}$$

For $x = -1$:

84

$$-\frac{(-1)}{3} + \frac{1}{2} - 2 = \frac{1}{3} + \frac{1}{2} - 2 = \frac{5}{6} - 2 = -\frac{7}{6}$$

Subtracting the lower limit evaluation from the upper limit evaluation:

$$A = \frac{10}{3} - \left(-\frac{7}{6}\right) = \frac{10}{3} + \frac{7}{6} = \frac{20}{6} + \frac{7}{6} = \frac{27}{6} = \frac{9}{2}$$

Therefore, the area enclosed between the curves $y = x^2$ and $y = x + 2$ is $\frac{9}{2}$ square units.

$$A = \frac{9}{2}$$

5.2 Compute the volume of the solid formed by revolving $y = \sqrt{x}$ around the x-axis from $x = 0$ to $x = 4$.

To compute the volume of the solid formed by revolving the curve $y = \sqrt{x}$ around the x-axis from $x = 0$ to $x = 4$, we will use the Disk Method. This method is appropriate for solids of revolution where the cross-sections perpendicular to the axis of rotation are disks (i.e., circles).

Disk Method Overview

The Disk Method involves integrating the area of circular disks along the axis of rotation. The volume V is given by:

$$V = \pi \int_a^b [f(x)]^2 \, dx$$

where:

- $f(x)$ is the function being revolved around the axis.

- $[f(x)]^2$ represents the area of a circular disk with radius $f(x)$.

- a and b are the bounds of integration along the axis of rotation.

Applying the Disk Method

85

Given:

$$y = \sqrt{x}$$

We are to revolve this curve around the x-axis from $x = 0$ to $x = 4$.

Step 1: Identify $f(x)$ and the bounds

Here,

$$f(x) = \sqrt{x}$$

and the bounds are $a = 0$ and $b = 4$.

Step 2: Set up the integral for volume

Using the Disk Method formula:

$$V = \pi \int_0^4 [\sqrt{x}]^2 \, dx$$

Simplify the integrand:

$$[\sqrt{x}]^2 = x$$

So, the integral becomes:

$$V = \pi \int_0^4 x \, dx$$

Step 3: Evaluate the integral

Compute the definite integral:

$$\int_0^4 x \, dx = \left[\frac{x^2}{2}\right]_0^4 = \frac{4^2}{2} - \frac{0^2}{2} = \frac{16}{2} - 0 = 8$$

Step 4: Multiply by π to find the volume

$$V = \pi \times 8 = 8\pi$$

The volume of the solid formed by revolving $y = \sqrt{x}$ around the x-axis from $x = 0$ to $x = 4$ is:

$$V = 8\pi \text{ cubic units}$$

5.3 Determine the arc length of $y = \ln(\cosh(x))$ from $x = 0$ to $x = 1$

To determine the arc length of the curve defined by $y = \ln(\cosh(x))$ over the interval $x \in [0, 1]$, we utilize the formula for the arc length of a function $y = f(x)$ from $x = a$ to $x = b$:

$$S = \int_a^b \sqrt{1 + \left(\frac{dy}{dx}\right)^2}\, dx$$

This integral calculates the cumulative length of the infinitesimal segments of the curve between $x = a$ and $x = b$.

Step 1: Compute the derivative $\frac{dy}{dx}$

Given:

$$y = \ln(\cosh(x))$$

First, we find the first derivative of y with respect to x:

$$\frac{dy}{dx} = \frac{d}{dx}\left[\ln(\cosh(x))\right]$$

Using the chain rule, where $\frac{d}{dx}\ln(u) = \frac{1}{u}\frac{du}{dx}$, we have:

$$\frac{dy}{dx} = \frac{1}{\cosh(x)} \cdot \frac{d}{dx}\cosh(x) = \frac{\sinh(x)}{\cosh(x)} = \tanh(x)$$

Thus:

$$\frac{dy}{dx} = \tanh(x)$$

Step 2: Formulate the integrand $\sqrt{1 + \left(\frac{dy}{dx}\right)^2}$

Substituting $\frac{dy}{dx}$ into the arc length formula:

$$S = \int_0^1 \sqrt{1 + \tanh^2(x)}\, dx$$

To simplify the integrand, we use the identity for hyperbolic functions:

87

$$\tanh^2(x) + \text{sech}^2(x) = 1$$

This implies:

$$\tanh^2(x) = 1 - \text{sech}^2(x)$$

Substituting back into the integrand:

$$\sqrt{1 + \tanh^2(x)} = \sqrt{1 + 1 - \text{sech}^2(x)} = \sqrt{2 - \text{sech}^2(x)}$$

Therefore, the arc length integral becomes:

$$S = \int_0^1 \sqrt{2 - \text{sech}^2(x)}\, dx$$

Step 3: Evaluate the integral

The integral:

$$S = \int_0^1 \sqrt{2 - \text{sech}^2(x)}\, dx$$

does not have an elementary antiderivative, meaning it cannot be expressed in terms of basic functions such as polynomials, exponentials, logarithms, trigonometric, or hypergeometric functions. Therefore, we evaluate it using numerical integration methods.

One common approach is to use the Simpson's Rule, which approximates the integral by dividing the interval into an even number of subintervals and fitting parabolas to segments of the curve.

Alternatively, we can use numerical integration tools or calculators capable of performing definite integrals numerically. For illustration, we will approximate the integral using Simpson's Rule with $n = 4$ subintervals.

Step 4: Apply Simpson's Rule

Simpson's Rule for n subintervals (where n is even) is given by:

$$\int_a^b f(x)\, dx \approx \frac{\Delta x}{3}\left[f(x_0) + 4\sum_{\text{odd } i} f(x_i) + 2\sum_{\text{even } i} f(x_i) + f(x_n) \right]$$

where $\Delta x = \frac{b-a}{n}$ and $x_i = a + i\Delta x$.

88

For $n = 4$, $a = 0$, $b = 1$:

$$\Delta x = \frac{1 - 0}{4} = 0.25$$

The points are $x_0 = 0$, $x_1 = 0.25$, $x_2 = 0.5$, $x_3 = 0.75$, $x_4 = 1$.

Compute $f(x) = \sqrt{2 - \text{sech}^2(x)}$ at each point:

$$f(0) = \sqrt{2 - \text{sech}^2(0)} = \sqrt{2 - 1} = 1$$
$$f(0.25) = \sqrt{2 - \text{sech}^2(0.25)} \approx \sqrt{2 - 0.967204} \approx \sqrt{1.032796} \approx 1.0162$$
$$f(0.5) = \sqrt{2 - \text{sech}^2(0.5)} \approx \sqrt{2 - 0.786447} \approx \sqrt{1.213553} \approx 1.1016$$
$$f(0.75) = \sqrt{2 - \text{sech}^2(0.75)} \approx \sqrt{2 - 0.649360} \approx \sqrt{1.350640} \approx 1.1624$$
$$f(1) = \sqrt{2 - \text{sech}^2(1)} \approx \sqrt{2 - 0.419974} \approx \sqrt{1.580026} \approx 1.2570$$

Now, apply Simpson's Rule:

$$S \approx \frac{0.25}{3} \left[1 + 4(1.0162 + 1.1624) + 2(1.1016) + 1.2570 \right]$$

Compute the sums:

$$4 \sum_{\text{odd } i} f(x_i) = 4(f(0.25) + f(0.75))$$
$$= 4(1.0162 + 1.1624)$$
$$= 4(2.1786)$$
$$= 8.7144$$
$$2 \sum_{\text{even } i} f(x_i) = 2(f(0.5))$$
$$= 2(1.1016)$$
$$= 2.2032$$

Substitute back:

$$S \approx \frac{0.25}{3} \left[1 + 8.7144 + 2.2032 + 1.2570\right]$$

$$= \frac{0.25}{3} \left[13.1746\right]$$

$$\approx \frac{0.25}{3} \times 13.1746$$

$$\approx 0.0833 \times 13.1746$$

$$\approx 1.0979$$

Step 5: Conclusion

Using Simpson's Rule with $n = 4$ subintervals, we approximate the arc length of the curve $y = \ln(\cosh(x))$ from $x = 0$ to $x = 1$ as:

$$S \approx 1.098 \, \text{units}$$

For higher accuracy, an increased number of subintervals or more advanced numerical integration techniques can be employed.

5.4 Calculate the surface area of revolution for $y = e^x$ rotated about the x-axis from $x = 0$ to $x = 1$.

To find the surface area of the curve $y = e^x$ rotated about the x-axis from $x = 0$ to $x = 1$, we use the formula for the surface area S of a surface of revolution around the x-axis:

$$S = 2\pi \int_a^b y \sqrt{1 + \left(\frac{dy}{dx}\right)^2} \, dx$$

Given:
$$y = e^x$$
$$a = 0,$$
$$b = 1$$

• **Step 1: Compute $\frac{dy}{dx}$**

$$\frac{dy}{dx} = \frac{d}{dx} e^x = e^x$$

90

- **Step 2: Substitute y and $\frac{dy}{dx}$ into the surface area formula**

$$S = 2\pi \int_0^1 e^x \sqrt{1 + (e^x)^2}\, dx = 2\pi \int_0^1 e^x \sqrt{1 + e^{2x}}\, dx$$

- **Step 3: Make a substitution to simplify the integral** Let:

$$u = e^x \quad \Rightarrow \quad du = e^x\, dx \quad \Rightarrow \quad dx = \frac{du}{u}$$

When $x = 0$, $u = e^0 = 1$. When $x = 1$, $u = e^1 = e$.

Substituting into the integral:

$$S = 2\pi \int_{u=1}^{u=e} u \cdot \sqrt{1 + u^2} \cdot \frac{du}{u} = 2\pi \int_1^e \sqrt{1 + u^2}\, du$$

- **Step 4: Evaluate the integral** $\int \sqrt{1 + u^2}\, du$ This is a standard integral that can be solved using hyperbolic substitution or by recognizing it as a form suitable for integration by parts. The integral is:

$$\int \sqrt{1 + u^2}\, du = \frac{1}{2}\left(u\sqrt{1 + u^2} + \sinh^{-1}(u) \right) + C$$

Alternatively, using logarithmic expression:

$$\int \sqrt{1 + u^2}\, du = \frac{1}{2}\left(u\sqrt{1 + u^2} + \ln\left(u + \sqrt{1 + u^2} \right) \right) + C$$

- **Step 5: Apply the limits of integration**

$$S = 2\pi \left[\frac{1}{2}\left(u\sqrt{1 + u^2} + \ln\left(u + \sqrt{1 + u^2} \right) \right) \right]_1^e$$

Simplifying the constants:

$$S = \pi \left[u\sqrt{1 + u^2} + \ln\left(u + \sqrt{1 + u^2} \right) \right]_1^e$$

- **Step 6: Evaluate at the upper and lower limits**

$$S = \pi \left(e\sqrt{1 + e^2} + \ln\left(e + \sqrt{1 + e^2} \right) \right) - \pi \left(1 \cdot \sqrt{1 + 1^2} + \ln\left(1 + \sqrt{1 + 1^2} \right) \right)$$

Simplify each term:

$$\sqrt{1 + e^2} = \sqrt{e^2 + 1}$$
$$\sqrt{1 + 1^2} = \sqrt{2}$$

Thus,

$$S = \pi \left(e\sqrt{e^2 + 1} + \ln\left(e + \sqrt{e^2 + 1} \right) \right) - \pi \left(\sqrt{2} + \ln\left(1 + \sqrt{2} \right) \right)$$

91

- **Final Expression:**

$$S = \pi \left(e\sqrt{e^2 + 1} - \sqrt{2} \right) + \pi \left(\ln \left(e + \sqrt{e^2 + 1} \right) - \ln \left(1 + \sqrt{2} \right) \right)$$

This expression represents the exact surface area of the curve $y = e^x$ rotated about the x-axis from $x = 0$ to $x = 1$.

- **Numerical Approximation (Optional)** For a numerical approximation, we can calculate each term using the approximate value of $e \approx 2.71828$:

$$\sqrt{e^2 + 1} = \sqrt{7.3891 + 1} = \sqrt{8.3891} \approx 2.8963$$

$$\ln(e + \sqrt{e^2 + 1}) = \ln(2.71828 + 2.8963) = \ln(5.6146) \approx 1.725$$

$$\sqrt{2} \approx 1.4142$$

$$\ln(1 + \sqrt{2}) = \ln(2.4142) \approx 0.8814$$

Substituting these values:

$$S \approx \pi \left(2.71828 \times 2.8963 - 1.4142 \right) + \pi \left(1.725 - 0.8814 \right)$$

$$S \approx \pi \left(7.8633 - 1.4142 \right) + \pi \left(0.8436 \right)$$

$$S \approx \pi \times 6.4491 + \pi \times 0.8436 = \pi \times 7.2927 \approx 22.883$$

Therefore, the approximate surface area is:

$$S \approx 22.883 \text{ square units}$$

The exact surface area of revolution for $y = e^x$ rotated about the x-axis from $x = 0$ to $x = 1$ is:

$$S = \pi \left(e\sqrt{e^2 + 1} + \ln \left(e + \sqrt{e^2 + 1} \right) - \sqrt{2} - \ln \left(1 + \sqrt{2} \right) \right)$$

Which approximates to:

$$S \approx 22.883 \text{ square units}$$

5.5 Find the work done in pumping water out of a hemispherical tank

To determine the work required to pump water out of a hemispherical tank, we will use the principles of calculus, specifically the application of integrals to calculate the total work done against gravity.

Problem Setup

Consider a hemispherical tank with radius R filled with water. We aim to find the total work W required to pump all the water out of the tank to the top edge.

Coordinate System and Variables

- Let the origin of our coordinate system be at the bottom of the hemisphere.

- The vertical axis will be denoted by y, with $y = 0$ at the bottom and $y = R$ at the top of the hemisphere.

- Consider a thin horizontal slice of water at a height y with a small thickness dy.

Volume of a Thin Slice

The volume dV of the thin slice can be expressed in terms of its cross-sectional area and thickness:

$$dV = \pi x^2 \, dy$$

where x is the radius of the circular slice at height y.

From the geometry of the hemisphere, we have the relationship:

$$x^2 + y^2 = R^2 \quad \Rightarrow \quad x = \sqrt{R^2 - y^2}$$

Thus, the volume becomes:

$$dV = \pi(R^2 - y^2) \, dy$$

Weight of the Thin Slice

Let γ be the weight density of water (weight per unit volume). The weight dW of the thin slice is:

$$dW = \gamma \, dV = \gamma \pi (R^2 - y^2) \, dy$$

Distance to Pump the Water

Each slice of water at height y must be lifted a distance equal to the height difference between the slice and the top of the tank. Since the top is at $y = R$, the distance d each slice must be lifted is:

$$d = R - y$$

Work Done to Pump the Slice

The work dW_{work} required to pump the thin slice is the product of the weight of the slice and the distance it must be lifted:

$$dW_{\text{work}} = dW \cdot d = \gamma\pi(R^2 - y^2)(R - y)\,dy$$

Total Work

To find the total work W required to pump all the water out of the tank, we integrate dW_{work} from the bottom of the tank ($y = 0$) to the top ($y = R$):

$$W = \int_0^R \gamma\pi(R^2 - y^2)(R - y)\,dy$$

Expanding the Integrand

Before integrating, let's expand the integrand:

$$(R^2 - y^2)(R - y) = R(R^2 - y^2) - y(R^2 - y^2) = R^3 - Ry^2 - R^2y + y^3$$

Thus, the integral becomes:

$$W = \gamma\pi \int_0^R \left(R^3 - Ry^2 - R^2y + y^3\right)dy$$

Integrating Term by Term

Now, integrate each term separately:

$$\int_0^R R^3\,dy = R^3 y \Big|_0^R = R^4$$

$$\int_0^R Ry^2\,dy = R \cdot \frac{y^3}{3}\Big|_0^R = \frac{R^4}{3}$$

$$\int_0^R R^2 y\,dy = R^2 \cdot \frac{y^2}{2}\Big|_0^R = \frac{R^4}{2}$$

$$\int_0^R y^3\,dy = \frac{y^4}{4}\Big|_0^R = \frac{R^4}{4}$$

Substituting back into the expression for W:

$$W = \gamma\pi\left(R^4 - \frac{R^4}{3} - \frac{R^4}{2} + \frac{R^4}{4}\right)$$

Combining Like Terms

To simplify, find a common denominator (12) for the fractions:

$$W = \gamma \pi R^4 \left(\frac{12}{12} - \frac{4}{12} - \frac{6}{12} + \frac{3}{12} \right) = \gamma \pi R^4 \left(\frac{5}{12} \right)$$

Final Expression for Work

Therefore, the total work done in pumping the water out of the hemispherical tank is:

$$W = \frac{5}{12} \gamma \pi R^4$$

The work required to pump all the water out of a hemispherical tank of radius R is directly proportional to R^4, the weight density γ, and the constant factors arising from the integral calculation. This result illustrates the practical application of integral calculus in solving real-world physics and engineering problems related to fluid mechanics.

5.6 Compute the Center of Mass of a Lamina with Density $\delta(x) = x$ over $[0, 2]$

To determine the center of mass of a lamina (a thin, flat object) with a given density function, we use the concepts of mass and moments in calculus. Here, the lamina lies along the x-axis from $x = 0$ to $x = 2$, and the density at any point x is given by $\delta(x) = x$.

- **Mass (M):** The total mass of the lamina is obtained by integrating the density function over the given interval.

$$M = \int_a^b \delta(x)\, dx$$

- **First Moment about the Origin (M_x):** This is the integral of the product of the position x and the density function.

$$M_x = \int_a^b x \cdot \delta(x)\, dx$$

- **Center of Mass ($\bar{x}$):** The x-coordinate of the center of mass is the ratio of the first moment to the total mass.

$$\bar{x} = \frac{M_x}{M}$$

95

Calculating the Mass (M)

Given:

$$\delta(x) = x \quad \text{and} \quad a = 0,\, b = 2$$

Then,

$$M = \int_0^2 x\, dx$$

Evaluating the integral:

$$M = \left[\frac{x^2}{2}\right]_0^2 = \frac{2^2}{2} - \frac{0^2}{2} = \frac{4}{2} - 0 = 2$$

Calculating the First Moment (M_x)

$$M_x = \int_0^2 x \cdot x\, dx = \int_0^2 x^2\, dx$$

Evaluating the integral:

$$M_x = \left[\frac{x^3}{3}\right]_0^2 = \frac{2^3}{3} - \frac{0^3}{3} = \frac{8}{3} - 0 = \frac{8}{3}$$

Determining the Center of Mass ($\overline{x}$)

$$\overline{x} = \frac{M_x}{M} = \frac{\frac{8}{3}}{2} = \frac{8}{3} \times \frac{1}{2} = \frac{4}{3}$$

The center of mass of the lamina with density $\delta(x) = x$ over the interval $[0, 2]$ is located at:

$$\boxed{\overline{x} = \frac{4}{3}}$$

5.7 Determine the pressure on a submerged plate in a fluid

To determine the pressure on a submerged plate in a fluid, we will analyze the pressure distribution acting on the plate due to the fluid's hydrostatic properties. We will consider a vertical submerged plate for simplicity, but the methodology can be extended to plates of different orientations.

Hydrostatic Pressure Fundamentals

In a fluid at rest, the pressure at any point depends only on the depth of the point below the surface of the fluid. The fundamental relationship governing hydrostatic pressure is given by:

$$p = p_0 + \rho g h$$

where:

- p is the pressure at depth h,

- p_0 is the atmospheric pressure on the fluid surface,

- ρ is the density of the fluid,

- g is the acceleration due to gravity,

- h is the depth below the fluid surface.

Pressure Distribution on the Submerged Plate

Consider a vertical plate submerged in a fluid. Let the plate have a height H and width W. We will analyze the pressure distribution along the height of the plate.

Coordinate System Setup

Set up a coordinate system where:

- The y-axis is vertical, pointing upwards.

- The origin ($y = 0$) is at the fluid surface.

- The plate extends from $y = 0$ (top of the plate) to $y = H$ (bottom of the plate).

Pressure at a Differential Element

Consider a differential horizontal strip of the plate at a depth y with a small height Δy and width W. The pressure at this depth is:

$$p(y) = p_0 + \rho g y$$

Differential Force on the Element

97

The differential force ΔF exerted by the fluid on this strip is the product of the pressure at depth y and the area of the strip:

$$\Delta F = p(y) \cdot \Delta A = (p_0 + \rho gy) \cdot (W \cdot \Delta y)$$

Total Force on the Submerged Plate

To find the total force F exerted by the fluid on the entire plate, integrate the differential force over the height of the plate from $y = 0$ to $y = H$:

$$F = \int_0^H (p_0 + \rho gy)W \, dy$$

Evaluating the Integral

Separate the integral into two parts:

$$F = W \int_0^H p_0 \, dy + W \int_0^H \rho gy \, dy$$

Compute each integral:

$$\int_0^H p_0 \, dy = p_0 \int_0^H dy = p_0 H$$
$$\int_0^H \rho gy \, dy = \rho g \int_0^H y \, dy = \rho g \left[\frac{y^2}{2} \right]_0^H = \frac{1}{2} \rho g H^2$$

Total Force Expression

Combine the results:

$$F = W \left(p_0 H + \frac{1}{2} \rho g H^2 \right)$$

Alternatively, the total force can be expressed as:

$$F = p_0 A + \frac{1}{2} \rho g H^2 W$$

where $A = WH$ is the area of the plate.

Center of Pressure

98

The point where the total force can be considered to act is known as the center of pressure. To find the depth y_{cp} at which the total force acts, use the moment equilibrium about the surface:

$$y_{cp} = \frac{\int_0^H y \cdot p(y) W \, dy}{F}$$

Substitute $p(y) = p_0 + \rho g y$:

$$y_{cp} = \frac{W \int_0^H y(p_0 + \rho g y) \, dy}{W(p_0 H + \frac{1}{2}\rho g H^2)}$$

Simplify the expression:

$$y_{cp} = \frac{\int_0^H (p_0 y + \rho g y^2) \, dy}{p_0 H + \frac{1}{2}\rho g H^2}$$

$$= \frac{p_0 \left[\frac{y^2}{2}\right]_0^H + \rho g \left[\frac{y^3}{3}\right]_0^H}{p_0 H + \frac{1}{2}\rho g H^2}$$

$$= \frac{p_0 \frac{H^2}{2} + \rho g \frac{H^3}{3}}{p_0 H + \frac{1}{2}\rho g H^2}$$

$$= \frac{\frac{1}{2}p_0 H^2 + \frac{1}{3}\rho g H^3}{p_0 H + \frac{1}{2}\rho g H^2}$$

Factor H from numerator and denominator:

$$y_{cp} = \frac{H\left(\frac{1}{2}p_0 H + \frac{1}{3}\rho g H^2\right)}{H\left(p_0 + \frac{1}{2}\rho g H\right)}$$

$$= \frac{\frac{1}{2}p_0 H + \frac{1}{3}\rho g H^2}{p_0 + \frac{1}{2}\rho g H}$$

This expression gives the depth at which the resultant pressure force acts on the submerged plate.

The hydrostatic pressure on a submerged vertical plate increases linearly with depth. The total force exerted by the fluid on the plate is given by:

99

$$F = W \left(p_0 H + \frac{1}{2}\rho g H^2 \right)$$

This force acts at a depth known as the center of pressure, which is located below the centroid of the plate due to the increasing pressure with depth. The calculation of the center of pressure is essential for understanding the torque and stability related to submerged structures in engineering applications.

Example Calculation

Problem: A vertical rectangular plate of width $2\,\text{m}$ and height $3\,\text{m}$ is submerged vertically in water ($\rho = 1000\,\text{kg/m}^3$). The top of the plate is $1\,\text{m}$ below the water surface. Calculate the total force exerted by the water on the plate and determine the center of pressure.

Solution:

First, identify the given values:

- Width, $W = 2\,\text{m}$

- Height, $H = 3\,\text{m}$

- Density of water, $\rho = 1000\,\text{kg/m}^3$

- Acceleration due to gravity, $g = 9.81\,\text{m/s}^2$

- Atmospheric pressure, $p_0 = 101325\,\text{Pa}$ (if needed)

Total Force:

$$F = W \left(p_0 H + \frac{1}{2}\rho g H^2 \right)$$

Since the plate is submerged such that the top is $1\,\text{m}$ below the surface, the depth h for the pressure calculation varies from $1\,\text{m}$ to $4\,\text{m}$. However, assuming p_0 is atmospheric pressure and considering gauge pressure, we can set $p_0 = 0$ for simplifying the hydrostatic force calculation.

Thus,

$$F = W \left(\frac{1}{2}\rho g H^2 \right) = 2 \left(\frac{1}{2} \times 1000 \times 9.81 \times 3^2 \right)$$

100

$$F = 2 \left(\frac{1}{2} \times 1000 \times 9.81 \times 9 \right)$$
$$= 2 \times (500 \times 9.81 \times 9)$$
$$= 2 \times (500 \times 88.29)$$
$$= 2 \times 44145$$
$$= 88290 \, \text{N}$$

Center of Pressure:

$$y_{cp} = \frac{\frac{1}{2} p_0 H + \frac{1}{3} \rho g H^2}{p_0 + \frac{1}{2} \rho g H}$$

With $p_0 = 0$,

$$y_{cp} = \frac{\frac{1}{3} \rho g H^2}{\frac{1}{2} \rho g H}$$
$$= \frac{2}{3} H$$
$$= \frac{2}{3} \times 3$$
$$= 2 \, \text{m}$$

Therefore, the center of pressure is located 2 m below the top of the plate.

The pressure on a submerged plate in a fluid increases with depth due to hydrostatic pressure. By integrating the pressure distribution over the area of the plate, we can determine the total force exerted by the fluid and its point of action, the center of pressure. These calculations are fundamental in engineering applications involving submerged structures, such as dams, underwater vessels, and hydraulic systems.

5.8 Calculate the moment of inertia of a thin rod about its end

To calculate the moment of inertia of a thin rod about one of its ends, we will consider the following parameters and steps:

Given:

- Length of the rod, L

- Mass of the rod, M

Objective: Find the moment of inertia, I, of the rod about an axis perpendicular to the rod and passing through one end.

Understanding the Moment of Inertia

The moment of inertia is a measure of an object's resistance to rotational acceleration about a specific axis. For a continuous object, it is defined by the integral:

$$I = \int r^2 \, dm$$

where:

- r is the distance from the axis of rotation to an infinitesimal mass element dm.

- The integral is taken over the entire mass of the object.

Setting Up the Problem

1. Coordinate System:

    - Place the rod along the x-axis with one end at the origin ($x = 0$) and the other end at $x = L$.

    - The axis of rotation is perpendicular to the rod and passes through $x = 0$.

2. Linear Mass Density:

    - The mass per unit length (λ) of the rod is:

$$\lambda = \frac{M}{L}$$

3. Infinitesimal Mass Element:

   - Consider an infinitesimal segment of the rod of length dx at a distance x from the axis.
   - The mass of this segment is:

$$dm = \lambda\, dx = \frac{M}{L}\, dx$$

Calculating the Moment of Inertia

Substitute the expressions for dm and r into the integral for I:

$$I = \int_0^L x^2\, dm = \int_0^L x^2 \left(\frac{M}{L}\right) dx$$

$$I = \frac{M}{L} \int_0^L x^2\, dx$$

Evaluate the integral:

$$\int_0^L x^2\, dx = \left[\frac{x^3}{3}\right]_0^L = \frac{L^3}{3} - 0 = \frac{L^3}{3}$$

Substitute back into the expression for I:

$$I = \frac{M}{L} \cdot \frac{L^3}{3} = \frac{ML^2}{3}$$

Final Result

The moment of inertia of a thin rod of length L and mass M about an axis perpendicular to the rod and passing through one end is:

$$I = \frac{1}{3}ML^2$$

This derivation shows that the moment of inertia depends on both the mass of the rod and the square of its length. The result $I = \frac{1}{3}ML^2$ is fundamental in dynamics, especially in problems involving rotational motion of slender bodies.

103

5.9 Find the probability that a randomly chosen point falls under a given curve

To determine the probability that a randomly chosen point falls under a given curve, we need to consider the area under the curve relative to the total area of the region within which the point is selected.

Problem Statement

Let $f(x)$ be a continuous non-negative function defined on the interval $[a, b]$. We aim to find the probability that a randomly chosen point (x, y) within the rectangle defined by $a \leq x \leq b$ and $0 \leq y \leq M$ (where M is an upper bound for $f(x)$ on $[a, b]$) falls under the curve $y = f(x)$.

Assumptions

- The point (x, y) is chosen uniformly at random within the rectangle $[a, b] \times [0, M]$.

- The function $f(x)$ is continuous and non-negative on $[a, b]$.

- M is chosen such that $M \geq f(x)$ for all x in $[a, b]$.

Solution

The probability P that the randomly chosen point (x, y) lies under the curve $y = f(x)$ is given by the ratio of the area under the curve to the total area of the rectangle.

$$P = \frac{\text{Area under the curve}}{\text{Total area of the rectangle}}$$

Calculating the Area under the Curve

The area under the curve $y = f(x)$ from $x = a$ to $x = b$ is given by the definite integral:

$$\text{Area under the curve} = \int_a^b f(x)\, dx$$

Calculating the Total Area of the Rectangle

The total area $A_{\text{rectangle}}$ of the rectangle $[a, b] \times [0, M]$ is:

$$A_{\text{rectangle}} = (b - a) \times M$$

Expressing the Probability

Substituting the expressions for the areas into the probability formula:

$$P = \frac{\int_a^b f(x)\, dx}{(b-a) \times M}$$

Example

Consider $f(x) = x$ on the interval $[0, 1]$.

Calculating the Area under the Curve

$$\int_0^1 x\, dx = \left[\frac{1}{2}x^2\right]_0^1 = \frac{1}{2}$$

Calculating the Total Area of the Rectangle

Assuming $M = 1$ (since $f(x) = x \leq 1$ on $[0, 1]$),

$$A_{\text{rectangle}} = (1-0) \times 1 = 1$$

Calculating the Probability

$$P = \frac{\frac{1}{2}}{1} = \frac{1}{2}$$

Therefore, the probability that a randomly chosen point (x, y) within the rectangle $[0, 1] \times [0, 1]$ falls under the curve $y = x$ is $\frac{1}{2}$.

Generalization

The approach demonstrated in the example can be generalized to any continuous non-negative function $f(x)$ over an interval $[a, b]$. By calculating the definite integral of $f(x)$ over $[a, b]$ and dividing it by the area of the encompassing rectangle, we obtain the probability that a randomly selected point within the rectangle lies under the curve $y = f(x)$.

$$P = \frac{\int_a^b f(x)\, dx}{(b-a) \times M}$$

This method is fundamental in probability theory and has applications in various fields such as statistics, physics, and engineering, where determining the likelihood of an event within a continuous domain is required.

5.10 Evaluate the integral representing the electrical charge over a region

To evaluate the electrical charge over a given region, we start by understanding the relationship between charge density and total charge. The electrical charge Q in a region can be determined by integrating the charge density ρ over the specified volume V.

- Evaluate the electrical charge Q in the region bounded by the cylinder $x^2 + y^2 = a^2$ and between the planes $z = 0$ and $z = h$, where the charge density is given by $\rho(x, y, z) = \rho_0$, a constant.

Solution

To find the total charge Q, we use the integral:

$$Q = \iiint_V \rho(x, y, z)\, dV \tag{5.1}$$

Given that $\rho(x, y, z) = \rho_0$ is constant, the integral simplifies to:

$$Q = \rho_0 \iiint_V dV = \rho_0 \cdot \text{Volume of } V \tag{5.2}$$

However, to demonstrate the integration process, we will evaluate the integral in cylindrical coordinates, which is suitable for the given cylindrical symmetry.

Cylindrical Coordinates

In cylindrical coordinates, the volume element dV is:

$$dV = r\, dr\, d\theta\, dz \tag{5.3}$$

The limits for the cylindrical region are:

$$r \in [0, a]$$
$$\theta \in [0, 2\pi]$$
$$z \in [0, h]$$

Setting Up the Integral

106

Substituting $\rho(x, y, z) = \rho_0$ and dV into the expression for Q:

$$Q = \rho_0 \int_0^{2\pi} \int_0^a \int_0^h r \, dz \, dr \, d\theta \tag{5.4}$$

Evaluating the Integral

We evaluate the integral step by step, starting with the innermost integral.

$$
\begin{aligned}
Q &= \rho_0 \int_0^{2\pi} \int_0^a \left(\int_0^h dz \right) r \, dr \, d\theta \\
&= \rho_0 \int_0^{2\pi} \int_0^a h \cdot r \, dr \, d\theta \\
&= \rho_0 h \int_0^{2\pi} \left(\int_0^a r \, dr \right) d\theta \\
&= \rho_0 h \int_0^{2\pi} \left[\frac{1}{2} r^2 \right]_0^a d\theta \\
&= \rho_0 h \int_0^{2\pi} \left(\frac{1}{2} a^2 \right) d\theta \\
&= \rho_0 h \cdot \frac{1}{2} a^2 \int_0^{2\pi} d\theta \\
&= \rho_0 h \cdot \frac{1}{2} a^2 \cdot 2\pi \\
&= \rho_0 h a^2 \pi
\end{aligned}
$$

Final Answer

The total electrical charge Q over the specified cylindrical region is:

$$Q = \rho_0 \pi a^2 h \tag{5.5}$$

By setting up the integral in cylindrical coordinates and evaluating it step by step, we have determined that the total charge Q within the cylinder of radius a and height h with uniform charge density ρ_0 is $Q = \rho_0 \pi a^2 h$. This example illustrates the application of multiple integrals in calculating physical quantities such as electrical charge in a defined region.

Chapter 6

Sequences and Series

This chapter addresses the foundational concepts and properties of sequences and series, presenting problems that underscore their significance in calculus. It delves into convergence and divergence, tests for convergence, power series, and Taylor and Maclaurin series. Through these problems, readers will refine their analytical skills and develop a deep understanding of how sequences and series function within the broader scope of calculus. Detailed solutions accompany each problem, offering clarity and guidance to facilitate a comprehensive grasp of the material.

6.1 Determine if the sequence $a_n = \frac{n}{n+1}$ converges.

Solution:

To determine whether the sequence $a_n = \frac{n}{n+1}$ converges, we will analyze its behavior as n approaches infinity. A sequence $\{a_n\}$ is said to converge to a limit L if, for every $\epsilon > 0$, there exists a positive integer N such that for all $n \geq N$, the inequality $|a_n - L| < \epsilon$ holds. If such an L exists, the sequence converges to L; otherwise, it diverges.

Step 1: Express the Sequence

The given sequence is:

$$a_n = \frac{n}{n+1}$$

109

Step 2: Analyze the Limit as $n \to \infty$

We aim to find:

$$\lim_{n \to \infty} \frac{n}{n+1}$$

To evaluate this limit, we can divide both the numerator and the denominator by n, the highest power of n present in the expression:

$$\lim_{n \to \infty} \frac{n}{n+1} = \lim_{n \to \infty} \frac{\frac{n}{n}}{\frac{n}{n} + \frac{1}{n}} = \lim_{n \to \infty} \frac{1}{1 + \frac{1}{n}}$$

Step 3: Simplify the Expression

As n approaches infinity, $\frac{1}{n}$ approaches 0. Therefore:

$$\lim_{n \to \infty} \frac{1}{1 + \frac{1}{n}} = \frac{1}{1+0} = 1$$

Step 4: Conclusion

Since the limit of the sequence a_n as $n \to \infty$ exists and is equal to 1, we conclude that the sequence $a_n = \frac{n}{n+1}$ converges.

$$\boxed{\lim_{n \to \infty} \frac{n}{n+1} = 1}$$

Hence, the sequence $\{a_n\}$ converges to 1.

6.2 Test the series $\sum_{n=1}^{\infty} \frac{1}{n^2}$ for convergence.

To determine the convergence of the series

$$sum_{n=1}^{\infty} \frac{1}{n^2},$$

we can employ several convergence tests. In this solution, we will use the **p-series test** and the **Comparison Test** to establish the convergence of the given series.

- **p-Series Test**

A **p-series** is of the form

$$sum_{n=1}^{\infty} \frac{1}{n^p},$$

where p is a positive real number. The p-series test states that:

110

- If $p > 1$, the series converges.

- If $p \leq 1$, the series diverges.

In our case, the given series

$$sum_{n=1}^{\infty} \frac{1}{n^2}$$

is a p-series with $p = 2$.

Applying the p-Series Test:

$$p = 2 > 1 \implies \text{The series converges.}$$

- **Comparison Test**

The **Comparison Test** is another effective method to determine the convergence of a series. It involves comparing the given series to a known benchmark series.

Consider the series

$$sum_{n=1}^{\infty} \frac{1}{n^2},$$

and compare it to the series

$$sum_{n=1}^{\infty} \frac{1}{n^p}$$

with $p = 2$, which we already know is a convergent p-series.

Establishing the Comparison:

$$0 < \frac{1}{n^2} \leq \frac{1}{n^2} \quad \forall n \geq 1.$$

Since both series are identical and the benchmark series $\sum \frac{1}{n^2}$ is convergent, by the Comparison Test, the given series also converges.

Both the p-Series Test and the Comparison Test confirm that the series

$$sum_{n=1}^{\infty} \frac{1}{n^2}$$

is convergent.

- **Additional Remarks**

111

It's noteworthy to mention that the exact sum of this series is known and is given by:

$$sum_{n=1}^{\infty} \frac{1}{n^2} = \frac{\pi^2}{6}.$$

This result, known as the Basel problem, was famously solved by Leonhard Euler. While finding the exact sum is beyond the scope of convergence tests, it serves to illustrate the series' convergent nature with a finite sum.

Using the p-Series Test with $p = 2$ and the Comparison Test, we have rigorously established that the series $\sum_{n=1}^{\infty} \frac{1}{n^2}$ converges.

6.3 Find the sum of the series $\sum_{n=0}^{\infty} r^n$ for $|r| < 1$.

To find the sum of the infinite geometric series $\sum_{n=0}^{\infty} r^n$ where $|r| < 1$, we will follow a step-by-step approach:

- **1. Understanding the Series**

The given series is:

$$S = \sum_{n=0}^{\infty} r^n = 1 + r + r^2 + r^3 + r^4 + \cdots$$

This is an infinite geometric series where:

- The first term $a = 1$ (when $n = 0$).

- The common ratio r satisfies $|r| < 1$ to ensure convergence.

- **2. Convergence of the Series**

For an infinite geometric series $\sum_{n=0}^{\infty} ar^n$, the series converges if and only if $|r| < 1$. Since our series satisfies $|r| < 1$, it is convergent, and we can find its sum.

- **3. Deriving the Sum**

Let's denote the sum of the series by S:

$$S = 1 + r + r^2 + r^3 + r^4 + \cdots$$

Multiply both sides of the equation by r:

$$rS = r + r^2 + r^3 + r^4 + r^5 + \cdots$$

Now, subtract the second equation from the first:

$$S - rS = (1 + r + r^2 + r^3 + \cdots) - (r + r^2 + r^3 + r^4 + \cdots)$$

$$S(1 - r) = 1$$

Solving for S:

$$S = \frac{1}{1 - r}$$

- **4. Conclusion**

Therefore, the sum of the infinite geometric series $\sum_{n=0}^{\infty} r^n$ for $|r| < 1$ is:

$$\sum_{n=0}^{\infty} r^n = \frac{1}{1 - r}$$

- **5. Example**

Example: Find the sum of the series $\sum_{n=0}^{\infty} \left(\frac{1}{2}\right)^n$.

Solution: Here, $r = \frac{1}{2}$ and $|\frac{1}{2}| < 1$.

Applying the formula:

$$\sum_{n=0}^{\infty} \left(\frac{1}{2}\right)^n = \frac{1}{1 - \frac{1}{2}} = \frac{1}{\frac{1}{2}} = 2$$

Thus, the sum of the series is 2.

6.4 Determine if the series $\sum_{n=1}^{\infty} \frac{(-1)^{n+1}}{n}$ converges absolutely.

To determine whether the series

$$\sum_{n=1}^{\infty} \frac{(-1)^{n+1}}{n}$$

converges absolutely, we analyze the series of its absolute values and then consider the original series in the context of conditional convergence.

Absolute Convergence

A series $\sum_{n=1}^{\infty} a_n$ is said to *converge absolutely* if the series of absolute values $\sum_{n=1}^{\infty} |a_n|$ converges. Therefore, we first consider the series of absolute values of the given series:

$$\sum_{n=1}^{\infty} \left| \frac{(-1)^{n+1}}{n} \right| = \sum_{n=1}^{\infty} \frac{1}{n}$$

This is the well-known **harmonic series**.

Harmonic Series Divergence: It is a fundamental result in calculus that the harmonic series diverges. This can be shown using the **Integral Test**. Consider the integral

$$\int_1^{\infty} \frac{1}{x} \, dx = \lim_{t \to \infty} \int_1^t \frac{1}{x} \, dx = \lim_{t \to \infty} \ln t - \ln 1 = \infty$$

Since the integral diverges, by the Integral Test, the harmonic series also diverges:

$$\sum_{n=1}^{\infty} \frac{1}{n} = \infty$$

Therefore, the series of absolute values diverges, and hence the original series does **not** converge absolutely.

Conditional Convergence

Although the series does not converge absolutely, it may still converge conditionally. We examine the original series:

$$\sum_{n=1}^{\infty} \frac{(-1)^{n+1}}{n}$$

This is an **alternating series** of the form $\sum_{n=1}^{\infty} (-1)^{n+1} b_n$, where $b_n = \frac{1}{n}$.

To determine convergence, we apply the **Alternating Series Test**, which states that an alternating series $\sum_{n=1}^{\infty} (-1)^{n+1} b_n$ converges if the following two conditions are met:

1. The sequence $\{b_n\}$ is monotonically decreasing.

2. $\lim_{n \to \infty} b_n = 0$.

114

Verification of Conditions:

1. **Monotonicity:** The sequence $b_n = \frac{1}{n}$ is positive and decreases monotonically as n increases because $\frac{1}{n+1} < \frac{1}{n}$ for all $n \geq 1$.

2. **Limit to Zero:**

$$\lim_{n \to \infty} b_n = \lim_{n \to \infty} \frac{1}{n} = 0$$

Both conditions of the Alternating Series Test are satisfied.

Conclusion of Conditional Convergence: Since the alternating series satisfies the Alternating Series Test, the original series

$$\sum_{n=1}^{\infty} \frac{(-1)^{n+1}}{n}$$

converges.

Final Conclusion

The series

$$\sum_{n=1}^{\infty} \frac{(-1)^{n+1}}{n}$$

does not converge absolutely because the series of absolute values diverges. However, it does converge conditionally by the Alternating Series Test. Therefore, the series converges conditionally.

The series converges, but it does not converge absolutely.

6.5 Apply the Ratio Test to $\sum_{n=1}^{\infty} \frac{n!}{n^n}$

To determine the convergence of the series

$$\sum_{n=1}^{\infty} \frac{n!}{n^n},$$

we will apply the **Ratio Test**. The Ratio Test states that for a series $\sum_{n=1}^{\infty} a_n$, if

$$L = \lim_{n \to \infty} \left| \frac{a_{n+1}}{a_n} \right|,$$

then:

- If $L < 1$, the series **converges absolutely**.

115

- If $L > 1$, the series **diverges**.

- If $L = 1$, the test is **inconclusive**.

Let's apply this test to our series.

Step 1: Identify a_n

Given the series:

$$a_n = \frac{n!}{n^n}.$$

Step 2: Compute $\frac{a_{n+1}}{a_n}$

First, find an expression for a_{n+1}:

$$a_{n+1} = \frac{(n+1)!}{(n+1)^{n+1}}.$$

Now, compute the ratio:

$$\frac{a_{n+1}}{a_n} = \frac{(n+1)!}{(n+1)^{n+1}} \times \frac{n^n}{n!} = \frac{(n+1) \cdot n^n}{(n+1)^{n+1}} = \frac{n^n}{(n+1)^n}.$$

Step 3: Simplify the Ratio

Observe that:

$$\frac{n^n}{(n+1)^n} = \left(\frac{n}{n+1}\right)^n = \left(1 - \frac{1}{n+1}\right)^n.$$

Recall the limit:

$$\lim_{n \to \infty} \left(1 - \frac{1}{n+1}\right)^n = \lim_{n \to \infty} \left(\left(1 - \frac{1}{n+1}\right)^{n+1}\right)^{\frac{n}{n+1}} = e^{-1}.$$

Thus,

$$\lim_{n \to \infty} \frac{a_{n+1}}{a_n} = e^{-1} \approx 0.3679.$$

Step 4: Apply the Ratio Test

Since:

$$L = \lim_{n \to \infty} \left|\frac{a_{n+1}}{a_n}\right| = e^{-1} < 1,$$

by the Ratio Test, the series $\sum_{n=1}^{\infty} \frac{n!}{n^n}$ **converges absolutely**.

The application of the Ratio Test reveals that the series $\sum_{n=1}^{\infty} \frac{n!}{n^n}$ converges absolutely because the limit of the ratio of successive terms is less than one.

6.6   Use the Integral Test on $\sum_{n=2}^{\infty} \frac{1}{n \ln(n)}$.

To determine the convergence or divergence of the series

$$\sum_{n=2}^{\infty} \frac{1}{n \ln(n)},$$

we will apply the **Integral Test**. The Integral Test states that for a series of the form $\sum_{n=N}^{\infty} a_n$, where $a_n = f(n)$, if:

- $f(x)$ is continuous,

- $f(x)$ is positive, and

- $f(x)$ is decreasing for all $x \geq N$, then the series $\sum_{n=N}^{\infty} a_n$ and the improper integral $\int_{N}^{\infty} f(x)\, dx$ either both converge or both diverge.

Step 1: Define the Function $f(x)$

Let

$$f(x) = \frac{1}{x \ln(x)}$$

for $x \geq 2$.

Step 2: Verify the Conditions for the Integral Test

- **Continuity**: The function $f(x)$ is continuous for all $x > 1$ because the denominator $x \ln(x)$ is never zero in this domain.

- **Positivity**: For $x > 1$, both x and $\ln(x)$ are positive. Therefore, $f(x) > 0$ for all $x \geq 2$.

- **Monotonicity**: We need to show that $f(x)$ is decreasing for $x \geq 2$. To do this, we compute the derivative of $f(x)$:

$$f'(x) = \frac{d}{dx}\left(\frac{1}{x \ln(x)} \right) = \frac{-\ln(x) - 1}{x^2 [\ln(x)]^2}.$$

Since $x > 1$, both x^2 and $[\ln(x)]^2$ are positive, and $\ln(x) + 1 > 0$. Therefore, $f'(x) < 0$ for all $x \geq 2$, which means $f(x)$ is decreasing on this interval.

117

Step 3: Set Up the Improper Integral

We consider the improper integral:

$$\int_2^\infty \frac{1}{x \ln(x)} \, dx.$$

Step 4: Evaluate the Improper Integral

To evaluate the integral, we perform a substitution:

Let

$$u = \ln(x) \quad \Rightarrow \quad d\nu = \frac{1}{x} \, dx.$$

When $x = 2$, $\nu = \ln(2)$, and as $x \to \infty$, $\nu \to \infty$.

Substituting into the integral:

$$\int_2^\infty \frac{1}{x \ln(x)} \, dx = \int_{\ln(2)}^\infty \frac{1}{\nu} \, d\nu.$$

Now, we evaluate the integral:

$$\int_{\ln(2)}^\infty \frac{1}{\nu} \, d\nu = \lim_{b \to \infty} \int_{\ln(2)}^b \frac{1}{\nu} \, d\nu = \lim_{b \to \infty} \left[\ln |\nu| \right]_{\ln(2)}^b = \lim_{b \to \infty} \left(\ln(b) - \ln(\ln(2)) \right).$$

As $b \to \infty$, $\ln(b) \to \infty$. Therefore,

$$\int_2^\infty \frac{1}{x \ln(x)} \, dx = \infty.$$

Step 5: Apply the Integral Test

Since the improper integral

$$\int_2^\infty \frac{1}{x \ln(x)} \, dx$$

diverges to infinity, the Integral Test implies that the series

$$\sum_{n=2}^\infty \frac{1}{n \ln(n)}$$

also **diverges**.

By applying the Integral Test, we have shown that the series

$$\sum_{n=2}^\infty \frac{1}{n \ln(n)}$$

does not converge; instead, it diverges.

118

6.7 Find the radius of convergence for the power series $\sum_{n=0}^{\infty} c_n x^n$.

To determine the radius of convergence R for the power series $\sum_{n=0}^{\infty} c_n x^n$, we can employ two common methods: the **Ratio Test** and the **Root Test**. Both methods provide criteria to establish the value of R based on the behavior of the coefficients c_n as n approaches infinity.

- **Using the Ratio Test**

The Ratio Test examines the limit of the absolute value of the ratio of consecutive terms. Specifically, we consider: $L = \lim_{n \to \infty} \left| \frac{c_{n+1} x^{n+1}}{c_n x^n} \right|$. Simplifying the expression inside the limit, we have: $L = \lim_{n \to \infty} \left| \frac{c_{n+1}}{c_n} \right| \cdot |x|$. For the series to converge, the limit L must satisfy $L < 1$: $\lim_{n \to \infty} \left| \frac{c_{n+1}}{c_n} \right| \cdot |x| < 1$. Solving for $|x|$, we obtain: $|x| < \frac{1}{\lim_{n \to \infty} \left| \frac{c_{n+1}}{c_n} \right|}$. Thus, the radius of convergence R is given by: $R = \frac{1}{\lim_{n \to \infty} \left| \frac{c_{n+1}}{c_n} \right|}$. **Note**: If the limit L does not exist, or if it equals zero or infinity, the Ratio Test might be inconclusive or indicate that the series converges for all real numbers or diverges everywhere, respectively.

- **Using the Root Test**

The Root Test involves evaluating the limit superior of the n-th root of the absolute value of the terms of the series: $L = \limsup_{n \to \infty} \left(|c_n x^n| \right)^{1/n}$. This simplifies to: $L = |x| \cdot \limsup_{n \to \infty} |c_n|^{1/n}$. For convergence, we require $L < 1$: $|x| \cdot \limsup_{n \to \infty} |c_n|^{1/n} < 1$. Solving for $|x|$, we find: $|x| < \frac{1}{\limsup_{n \to \infty} |c_n|^{1/n}}$. Therefore, the radius of convergence R is: $R = \frac{1}{\limsup_{n \to \infty} |c_n|^{1/n}}$. **Interpretation**: The Root Test often provides a more straightforward computation for R, especially when dealing with coefficients that have a clear exponential growth or decay pattern.

Both the Ratio Test and the Root Test yield expressions for the radius of convergence R of the power series $\sum_{n=0}^{\infty} c_n x^n$. The choice of which test to use may depend on the nature of the coefficients c_n and the ease of evaluating the respective limits. Once R is determined, the interval of convergence can be expressed as: $|x| < R$, centered at the origin. To fully describe the interval of convergence, one must also examine the behavior of the series at the endpoints $x = \pm R$.

- **Example**

119

As an illustrative example, consider the power series $\sum_{n=0}^{\infty} \frac{x^n}{n!}$. Applying the Ratio Test: $L = \lim_{n\to\infty} \left| \frac{\frac{1}{(n+1)!}}{\frac{1}{n!}} \right| \cdot |x| = \lim_{n\to\infty} \frac{1}{n+1} \cdot |x| = 0 < 1$. Since $L = 0$ for all finite x, the radius of convergence R is infinity. Thus, the series converges for all real numbers x.

6.8 Compute the Taylor Series Expansion of e^x at $x = 0$.

To find the Taylor series expansion of the exponential function e^x centered at $x = 0$, also known as the Maclaurin series for e^x, we follow the standard procedure for deriving a Taylor series.

The Taylor series of a function $f(x)$ about $x = a$ is given by:

$$f(x) = \sum_{n=0}^{\infty} \frac{f^{(n)}(a)}{n!}(x - a)^n$$

where:

- $f^{(n)}(a)$ denotes the n-th derivative of f evaluated at $x = a$,

- $n!$ is the factorial of n,

- and the series is an infinite sum from $n = 0$ to $n = \infty$.

For the Maclaurin series, we set $a = 0$:

$$f(x) = \sum_{n=0}^{\infty} \frac{f^{(n)}(0)}{n!}x^n$$

Step 1: Compute the Derivatives of e^x

The exponential function has a unique and simple property where all its derivatives are equal to the function itself. Therefore, for all $n \geq 0$:

$$f^{(n)}(x) = \frac{d^n}{dx^n}e^x = e^x$$

Step 2: Evaluate the Derivatives at $x = 0$

Evaluating the n-th derivative at $x = 0$:

$$f^{(n)}(0) = e^0 = 1$$

Step 3: Substitute into the Taylor Series Formula

Substituting $f^{(n)}(0) = 1$ into the Maclaurin series formula:

$$e^x = \sum_{n=0}^{\infty} \frac{1}{n!} x^n$$

Final Taylor Series Expansion

Thus, the Taylor series expansion of e^x about $x = 0$ is:

$$e^x = \sum_{n=0}^{\infty} \frac{x^n}{n!} = 1 + x + \frac{x^2}{2!} + \frac{x^3}{3!} + \frac{x^4}{4!} + \cdots$$

Conclusion

The Taylor series for e^x converges for all real numbers x, providing an infinite polynomial representation of the exponential function. This series is fundamental in various applications of calculus, including solving differential equations, evaluating limits, and approximating functions.

$$e^x = \sum_{n=0}^{\infty} \frac{x^n}{n!}$$

6.9 Determine the nth Term of the Sequence Defined Recursively

Consider the sequence defined recursively by:

$$a_1 = 2, \quad a_{n+1} = 3a_n + 4 \quad \text{for } n \geq 1.$$

Our goal is to find an explicit formula for the nth term, a_n, of this sequence.

- The given recursive relation is linear and non-homogeneous:

$$a_{n+1} = 3a_n + 4.$$

To solve this, we will find the general solution by addressing both the homogeneous and particular parts of the recurrence.

121

- First, consider the homogeneous version of the recurrence:

$$a_{n+1}^{(h)} = 3a_n^{(h)}.$$

This is a simple linear homogeneous recurrence relation. Its characteristic equation is:

$$r = 3.$$

Thus, the general solution to the homogeneous equation is:

$$a_n^{(h)} = C \cdot 3^n,$$

where C is a constant to be determined by initial conditions.

- Next, we seek a particular solution, $a_n^{(p)}$, to the non-homogeneous recurrence:

$$a_{n+1}^{(p)} = 3a_n^{(p)} + 4.$$

Since the non-homogeneous term is a constant, we propose a constant particular solution:

$$a_n^{(p)} = A,$$

where A is a constant. Substitute this into the recurrence:

$$A = 3A + 4.$$

Solving for A:

$$A - 3A = 4 \quad \Rightarrow \quad -2A = 4 \quad \Rightarrow \quad A = -2.$$

Thus, the particular solution is:

$$a_n^{(p)} = -2.$$

- The general solution to the non-homogeneous recurrence is the sum of the homogeneous and particular solutions:

$$a_n = a_n^{(h)} + a_n^{(p)} = C \cdot 3^n - 2.$$

- We use the initial condition $a_1 = 2$ to determine the constant C:

$$a_1 = C \cdot 3^1 - 2 = 3C - 2.$$

Setting this equal to 2:

$$3C - 2 = 2 \quad \Rightarrow \quad 3C = 4 \quad \Rightarrow \quad C = \frac{4}{3}.$$

Therefore, the explicit formula for the nth term is:

$$a_n = \frac{4}{3} \cdot 3^n - 2 = 4 \cdot 3^{n-1} - 2.$$

- Simplifying the expression, we obtain:

$$a_n = 4 \cdot 3^{n-1} - 2.$$

This formula allows us to compute the nth term of the sequence directly without referring to previous terms.

- To ensure the correctness of our solution, let's verify it with the first few terms.

- For $n = 1$:
$$a_1 = 4 \cdot 3^0 - 2 = 4 \cdot 1 - 2 = 2,$$

which matches the initial condition.

- For $n = 2$:
$$a_2 = 4 \cdot 3^1 - 2 = 12 - 2 = 10.$$

Using the recursive relation:

$$a_2 = 3a_1 + 4 = 3 \cdot 2 + 4 = 10,$$

which is consistent.

- For $n = 3$:
$$a_3 = 4 \cdot 3^2 - 2 = 36 - 2 = 34.$$

Using the recursive relation:

$$a_3 = 3a_2 + 4 = 3 \cdot 10 + 4 = 34,$$

which also matches.

The derived formula correctly produces the terms of the sequence, confirming its validity.

- We have successfully determined the explicit nth term of the recursively defined sequence. The final formula is:

$$a_n = 4 \cdot 3^{n-1} - 2.$$

This allows for efficient computation of any term in the sequence without iterative recursion.

6.10 Test the alternating series $\sum_{n=1}^{\infty} \frac{(-1)^{n+1}}{n^p}$ for convergence.

To determine the convergence of the alternating series

$$\sum_{n=1}^{\infty} \frac{(-1)^{n+1}}{n^p},$$

we will employ the **Alternating Series Test** (Leibniz's Test) and analyze the **absolute convergence** of the series.

Alternating Series Test

The Alternating Series Test states that the series

$$\sum_{n=1}^{\infty} (-1)^{n+1} b_n$$

converges provided that:

1. $b_n > 0$ for all n,

2. $b_{n+1} \leq b_n$ for all n (i.e., $\{b_n\}$ is a non-increasing sequence),

3. $\lim_{n \to \infty} b_n = 0$.

In our given series, $b_n = \frac{1}{n^p}$. We will verify the three conditions:

1. Positivity

For all $n \geq 1$ and $p > 0$,

$$b_n = \frac{1}{n^p} > 0.$$

Hence, the first condition is satisfied.

2. Monotonicity

We need to show that $\{b_n\}$ is non-increasing. Consider b_{n+1} and b_n:

$$b_{n+1} = \frac{1}{(n+1)^p} \quad \text{and} \quad b_n = \frac{1}{n^p}.$$

Since $n + 1 > n$ and $p > 0$, it follows that

$$(n+1)^p > n^p \quad \Rightarrow \quad \frac{1}{(n+1)^p} < \frac{1}{n^p} \quad \Rightarrow \quad b_{n+1} < b_n.$$

Thus, $\{b_n\}$ is a decreasing sequence, satisfying the second condition.

3. Limit to Zero

We evaluate the limit:

$$\lim_{n \to \infty} b_n = \lim_{n \to \infty} \frac{1}{n^p} = 0,$$

for any $p > 0$. Therefore, the third condition is met.

Conclusion from Alternating Series Test

Since all three conditions of the Alternating Series Test are satisfied for $p > 0$, the series

$$\sum_{n=1}^{\infty} \frac{(-1)^{n+1}}{n^p}$$

converges for all $p > 0$.

Absolute Convergence

To determine whether the series converges absolutely, we examine the absolute series:

$$\sum_{n=1}^{\infty} \left| \frac{(-1)^{n+1}}{n^p} \right| = \sum_{n=1}^{\infty} \frac{1}{n^p}.$$

This is the well-known **p-series**, which converges if and only if $p > 1$.

Case 1: $p > 1$

For $p > 1$, the p-series converges:

$$\sum_{n=1}^{\infty} \frac{1}{n^p} \quad \text{converges.}$$

Thus, the original alternating series converges **absolutely** for $p > 1$.

125

Case 2: $0 < p \leq 1$

For $0 < p \leq 1$, the p-series diverges:

$$\sum_{n=1}^{\infty} \frac{1}{n^p} \quad \text{diverges.}$$

Since the absolute series diverges but the original alternating series converges (from the Alternating Series Test), the series converges **conditionally** for $0 < p \leq 1$.

Summary of Convergence

- For $p > 1$: The series $\sum_{n=1}^{\infty} \frac{(-1)^{n+1}}{n^p}$ **converges absolutely**.

- For $0 < p \leq 1$: The series $\sum_{n=1}^{\infty} \frac{(-1)^{n+1}}{n^p}$ **converges conditionally**.

The alternating series $\sum_{n=1}^{\infty} \frac{(-1)^{n+1}}{n^p}$ converges for all real numbers $p > 0$. It converges absolutely when $p > 1$ and converges conditionally when $0 < p \leq 1$.

Final Remarks

Understanding the distinction between absolute and conditional convergence is crucial in the study of infinite series. The Alternating Series Test provides a powerful tool for establishing convergence in cases where absolute convergence may not hold, thereby broadening the scope of series that can be effectively analyzed in calculus.

Chapter 7

Parametric and Polar Coordinates

This chapter explores the concepts and applications of parametric and polar coordinates, providing problems that highlight their utility in representing curves and complex geometric shapes. It covers topics such as parametrizing curves, converting between coordinate systems, analyzing polar equations, and calculating areas and lengths. By solving these problems, readers will gain proficiency in handling parametric and polar coordinates, expanding their understanding of different mathematical representations. Detailed solutions are included to facilitate comprehension and mastery of these topics.

7.1 Eliminate the parameter to find the Cartesian equation of $x = \cos(t), y = \sin(t)$.

To eliminate the parameter t and find a Cartesian equation that relates x and y, we follow these steps:

- **Given Parametric Equations**

$$x = \cos(t) \quad (1)$$
$$y = \sin(t) \quad (2)$$

127

- **Objective** Find a Cartesian equation of the form $F(x, y) = 0$ that represents the same curve as the given parametric equations, without involving the parameter t.

- **Step 1: Square Both Equations** First, we square both sides of equations (1) and (2) to utilize the Pythagorean identity.

$$x^2 = \cos^2(t) \quad (3)$$
$$y^2 = \sin^2(t) \quad (4)$$

- **Step 2: Add the Squared Equations** Next, we add equations (3) and (4) together:

$$x^2 + y^2 = \cos^2(t) + \sin^2(t)$$

- **Step 3: Apply the Pythagorean Identity** Recall the fundamental Pythagorean identity in trigonometry:

$$\cos^2(t) + \sin^2(t) = 1$$

Substituting this into our equation:

$$x^2 + y^2 = 1$$

The Cartesian equation that represents the same curve as the given parametric equations is:

$$x^2 + y^2 = 1$$

This equation represents a unit circle centered at the origin in the Cartesian coordinate system.

- **Verification** To verify that the parametric equations $x = \cos(t)$ and $y = \sin(t)$ satisfy the Cartesian equation $x^2 + y^2 = 1$, substitute x and y from the parametric equations into the Cartesian equation:

$$x^2 + y^2 = (\cos(t))^2 + (\sin(t))^2$$
$$= \cos^2(t) + \sin^2(t)$$
$$= 1 \quad \text{(by the Pythagorean identity)}$$

This confirms that all points (x, y) defined by the parametric equations lie on the unit circle $x^2 + y^2 = 1$.

- **Graphical Representation** The graph of the equation $x^2 + y^2 = 1$ is a circle with radius 1 centered at the origin $(0, 0)$. The parametric equations $x = \cos(t)$ and $y = \sin(t)$ trace this circle as the parameter t varies from 0 to 2π.

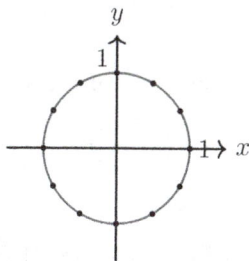

By eliminating the parameter t from the parametric equations $x = \cos(t)$ and $y = \sin(t)$, we derived the Cartesian equation $x^2 + y^2 = 1$, which represents a unit circle centered at the origin. This process involves using fundamental trigonometric identities and algebraic manipulation to transition from parametric to Cartesian forms.

7.2 Find $\frac{dy}{dx}$ for the parametric equations $x = t^2, y = t^3$.

To find the derivative $\frac{dy}{dx}$ for the given parametric equations:

$$x = t^2 \quad \text{and} \quad y = t^3,$$

we will use the chain rule for differentiation in parametric form. The derivative $\frac{dy}{dx}$ can be expressed as:

$$\frac{dy}{dx} = \frac{\frac{dy}{dt}}{\frac{dx}{dt}}.$$

- First, we find the derivatives of x and y with respect to the parameter t:

$$\frac{dx}{dt} = \frac{d}{dt}(t^2) = 2t,$$

$$\frac{dy}{dt} = \frac{d}{dt}(t^3) = 3t^2.$$

129

- Using the chain rule for parametric equations:

$$\frac{dy}{dx} = \frac{\frac{dy}{dt}}{\frac{dx}{dt}} = \frac{3t^2}{2t}.$$

- Simplify the fraction by canceling a factor of t in the numerator and the denominator:

$$\frac{dy}{dx} = \frac{3t^2}{2t} = \frac{3t}{2}.$$

Thus, the derivative $\frac{dy}{dx}$ for the given parametric equations is:

$$\frac{dy}{dx} = \frac{3}{2}t.$$

7.3 Compute the area enclosed by one loop of $r = \sin(2\theta)$.

To determine the area enclosed by one loop of the polar curve $r = \sin(2\theta)$, we will use the formula for the area A in polar coordinates:

$$A = \frac{1}{2} \int_\alpha^\beta r^2 \, d\theta$$

where α and β are the bounds of θ that define one complete loop of the curve.

Understanding the Curve

The given polar equation is:

$$r = \sin(2\theta)$$

This is known as a *rose curve*. The general form of a rose curve is $r = \sin(n\theta)$ or $r = \cos(n\theta)$, where:

- If n is an integer:
    - If n is odd, the rose has n petals.
    - If n is even, the rose has $2n$ petals.

In our case, $n = 2$, so the curve $r = \sin(2\theta)$ has $2 \times 2 = 4$ petals.

130

Determining the Bounds for One Petal

To find the area of one petal, we need to determine the interval $[\alpha, \beta]$ over which r traces out a single loop.

Set $r = 0$ to find the starting and ending angles:

$$\sin(2\theta) = 0 \implies 2\theta = k\pi \implies \theta = \frac{k\pi}{2}, \quad k \in \mathbb{Z}$$

Between $\theta = 0$ and $\theta = \frac{\pi}{2}$, $\sin(2\theta)$ goes from 0 to 1 and back to 0, tracing out one petal.

Thus, the bounds for one petal are:

$$\alpha = 0, \quad \beta = \frac{\pi}{2}$$

Calculating the Area

Substitute $r = \sin(2\theta)$ and the bounds into the area formula:

$$A = \frac{1}{2} \int_0^{\frac{\pi}{2}} \sin^2(2\theta) \, d\theta$$

To evaluate this integral, we use the power-reduction identity for sine:

$$\sin^2(x) = \frac{1 - \cos(2x)}{2}$$

Applying this identity:

$$A = \frac{1}{2} \int_0^{\frac{\pi}{2}} \frac{1 - \cos(4\theta)}{2} \, d\theta = \frac{1}{4} \int_0^{\frac{\pi}{2}} (1 - \cos(4\theta)) \, d\theta$$

Now, integrate term by term:

$$\int (1 - \cos(4\theta)) \, d\theta = \int 1 \, d\theta - \int \cos(4\theta) \, d\theta = \theta - \frac{\sin(4\theta)}{4} + C$$

Applying the limits of integration:

131

$$A = \frac{1}{4}\left[\theta - \frac{\sin(4\theta)}{4}\right]_0^{\frac{\pi}{2}} = \frac{1}{4}\left(\left(\frac{\pi}{2} - \frac{\sin(2\pi)}{4}\right) - \left(0 - \frac{\sin(0)}{4}\right)\right)$$

Since $\sin(2\pi) = \sin(0) = 0$:

$$A = \frac{1}{4}\left(\frac{\pi}{2} - 0 - 0 + 0\right) = \frac{1}{4} \times \frac{\pi}{2} = \frac{\pi}{8}$$

The area enclosed by one loop of the polar curve $r = \sin(2\theta)$ is:

$$A = \frac{\pi}{8}$$

7.4 Determine the arc length of $x = t, y = t^2$ from $t = 0$ to $t = 1$.

To determine the arc length of the parametric curve defined by $x = t$ and $y = t^2$ from $t = 0$ to $t = 1$, we will use the formula for the arc length of a parametric curve.

Arc Length Formula for Parametric Curves

The arc length L of a curve defined parametrically by $x = x(t)$ and $y = y(t)$ for t in the interval $[a, b]$ is given by:

$$L = \int_a^b \sqrt{\left(\frac{dx}{dt}\right)^2 + \left(\frac{dy}{dt}\right)^2}\, dt$$

Applying the Formula

Given:

$$x = t \quad \text{and} \quad y = t^2$$

First, compute the derivatives $\frac{dx}{dt}$ and $\frac{dy}{dt}$:

$$\frac{dx}{dt} = \frac{d}{dt}t = 1$$

132

$$\frac{dy}{dt} = \frac{d}{dt}t^2 = 2t$$

Substitute these derivatives into the arc length formula:

$$L = \int_0^1 \sqrt{(1)^2 + (2t)^2}\, dt = \int_0^1 \sqrt{1 + 4t^2}\, dt$$

Evaluating the Integral

To evaluate the integral $\int \sqrt{1 + 4t^2}\, dt$, we can use a trigonometric substitution or recognize it as a standard integral. Here, we'll use a standard integral approach.

Let's set:

$$u = 2t \quad \Rightarrow \quad du = 2\, dt \quad \Rightarrow \quad dt = \frac{du}{2}$$

When $t = 0$, $u = 0$; when $t = 1$, $u = 2$.

Substituting into the integral:

$$L = \int_0^1 \sqrt{1 + 4t^2}\, dt = \int_0^2 \sqrt{1 + u^2} \cdot \frac{du}{2} = \frac{1}{2} \int_0^2 \sqrt{1 + u^2}\, du$$

The integral $\int \sqrt{1 + u^2}\, du$ is a standard integral and can be expressed as:

$$\int \sqrt{1 + u^2}\, du = \frac{u}{2}\sqrt{1 + u^2} + \frac{1}{2}\sinh^{-1}(u) + C$$

Since $\sinh^{-1}(u) = \ln\left(u + \sqrt{1 + u^2}\right)$, we have:

$$\int \sqrt{1 + u^2}\, du = \frac{u}{2}\sqrt{1 + u^2} + \frac{1}{2}\ln\left(u + \sqrt{1 + u^2}\right) + C$$

Applying the limits of integration:

$$L = \frac{1}{2}\left[\frac{u}{2}\sqrt{1 + u^2} + \frac{1}{2}\ln\left(u + \sqrt{1 + u^2}\right)\right]_0^2$$

Evaluating at $u = 2$:

133

$$\frac{2}{2}\sqrt{1+2^2} + \frac{1}{2}\ln\left(2 + \sqrt{1+2^2}\right) = \sqrt{5} + \frac{1}{2}\ln\left(2 + \sqrt{5}\right)$$

Evaluating at $u = 0$:

$$0 + \frac{1}{2}\ln\left(0 + \sqrt{1+0^2}\right) = 0 + \frac{1}{2}\ln(1) = 0$$

Subtracting the lower limit from the upper limit:

$$L = \frac{1}{2}\left(\sqrt{5} + \frac{1}{2}\ln\left(2 + \sqrt{5}\right)\right) = \frac{\sqrt{5}}{2} + \frac{1}{4}\ln\left(2 + \sqrt{5}\right)$$

Final Result

Thus, the arc length L from $t = 0$ to $t = 1$ is:

$$L = \frac{\sqrt{5}}{2} + \frac{1}{4}\ln\left(2 + \sqrt{5}\right) \approx 1.47894$$

The arc length of the curve defined by $x = t$ and $y = t^2$ from $t = 0$ to $t = 1$ is exactly $\frac{\sqrt{5}}{2} + \frac{1}{4}\ln\left(2 + \sqrt{5}\right)$, which is approximately 1.47894 units.

7.5 Convert the polar equation $r = 2\sin(\theta)$ to Cartesian coordinates

To convert the polar equation $r = 2\sin(\theta)$ to Cartesian coordinates, we follow these systematic steps:

Recall the Relationships Between Polar and Cartesian Coordinates

In polar coordinates, the position of a point is determined by the radius r and the angle θ. The relationships between polar coordinates (r, θ) and Cartesian coordinates (x, y) are given by:

$$\begin{cases} x = r\cos(\theta) \\ y = r\sin(\theta) \\ r^2 = x^2 + y^2 \\ \tan(\theta) = \dfrac{y}{x} \end{cases}$$

Start with the Given Polar Equation

We begin with the polar equation:

$$r = 2\sin(\theta)$$

Express $\sin(\theta)$ in Terms of Cartesian Coordinates

From the relationship $y = r\sin(\theta)$, we can express $\sin(\theta)$ as:

$$\sin(\theta) = \frac{y}{r}$$

Substitute $\sin(\theta)$ into the Polar Equation

Substituting $\sin(\theta)$ into the original equation:

$$r = 2\left(\frac{y}{r}\right)$$

Solve for r

Multiply both sides of the equation by r to eliminate the denominator:

$$r^2 = 2y$$

Express r^2 in Terms of Cartesian Coordinates

Recall that $r^2 = x^2 + y^2$. Substitute this into the equation:

$$x^2 + y^2 = 2y$$

Rearrange the Equation to Standard Cartesian Form

Bring all terms to one side to set the equation to zero:

$$x^2 + y^2 - 2y = 0$$

Complete the Square for the y-Terms

To express the equation in standard form, complete the square for the y-terms:

$$x^2 + (y^2 - 2y) = 0$$
$$x^2 + \left(y^2 - 2y + 1\right) = 1$$
$$x^2 + (y - 1)^2 = 1$$

Interpret the Cartesian Equation

The final equation $x^2 + (y-1)^2 = 1$ represents a circle in Cartesian coordinates with:

135

- Center at $(0,1)$

- Radius 1

Thus, the polar equation $r = 2\sin(\theta)$ converted to Cartesian coordinates is:

$$x^2 + (y-1)^2 = 1$$

which describes a circle centered at $(0,1)$ with a radius of 1.

Graphical Representation

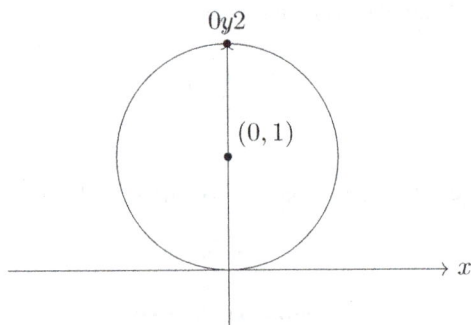

The graph above illustrates the circle $x^2 + (y-1)^2 = 1$, confirming the conversion from polar to Cartesian coordinates.

7.6 Find the area between the curves $r = 2$ and $r = 2\sin(\theta)$.

To determine the area enclosed between the polar curves $r = 2$ and $r = 2\sin(\theta)$, we will follow a systematic approach involving the following steps:

1. Graphical Analysis:

2. Determining the Points of Intersection:

3. Setting Up the Integral for the Area:

4. Calculating the Area:

- Graphical Analysis

136

First, let's understand the nature of the two curves:

- $r = 2$: This is a circle with a radius of 2 units centered at the origin.

- $r = 2\sin(\theta)$: This is a circle with a radius of 1 unit centered at $(0, 1)$ in Cartesian coordinates. It is known as the *cardioid*.

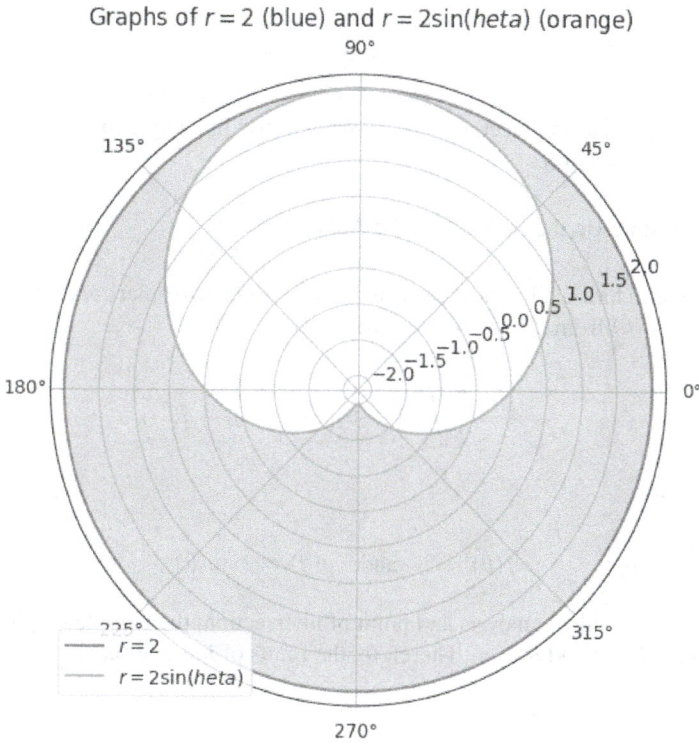

Graphs of $r = 2$ (blue) and $r = 2\sin(heta)$ (orange)

From the graph, it's evident that the curves intersect at certain angles, creating a bounded region whose area we aim to find.

- Determining the Points of Intersection

To find the points where the two curves intersect, we set $r = 2$ equal to $r = 2\sin(\theta)$:

$$2 = 2\sin(\theta)$$

Dividing both sides by 2:

$$1 = \sin(\theta)$$

Solving for θ:

$$\theta = \frac{\pi}{2}$$

Thus, the curves intersect at $\theta = \frac{\pi}{2}$. We will use this angle to determine the limits of integration.

- Setting Up the Integral for the Area

The general formula for the area A enclosed between two polar curves $r = f(\theta)$ and $r = g(\theta)$ from $\theta = \alpha$ to $\theta = \beta$ is:

$$A = \frac{1}{2} \int_{\alpha}^{\beta} \left([f(\theta)]^2 - [g(\theta)]^2 \right) d\theta$$

In our case:

$$f(\theta) = 2 \quad \text{and} \quad g(\theta) = 2\sin(\theta)$$

From the graphical analysis and point of intersection, the bounded region lies between $\theta = 0$ and $\theta = \pi$. Therefore, the limits of integration are $\alpha = 0$ and $\beta = \pi$.

- Calculating the Area

Substituting the values into the area formula:

$$A = \frac{1}{2} \int_{0}^{\pi} \left(2^2 - (2\sin(\theta))^2 \right) d\theta$$

Simplifying the integrand:

$$A = \frac{1}{2} \int_{0}^{\pi} \left(4 - 4\sin^2(\theta) \right) d\theta = 2 \int_{0}^{\pi} \left(1 - \sin^2(\theta) \right) d\theta$$

Recall the trigonometric identity:

$$\sin^2(\theta) = \frac{1 - \cos(2\theta)}{2}$$

Substituting this identity into the integral:

$$A = 2 \int_0^\pi \left(1 - \frac{1 - \cos(2\theta)}{2} \right) d\theta$$
$$= 2 \int_0^\pi \left(\frac{1 + \cos(2\theta)}{2} \right) d\theta$$
$$= \int_0^\pi (1 + \cos(2\theta)) \, d\theta$$

Now, integrate term by term:

$$A = \int_0^\pi 1 \, d\theta + \int_0^\pi \cos(2\theta) \, d\theta$$

Calculating each integral:

$$\int_0^\pi 1 \, d\theta = [\theta]_0^\pi = \pi - 0 = \pi$$

$$\int_0^\pi \cos(2\theta) \, d\theta = \left[\frac{\sin(2\theta)}{2} \right]_0^\pi = \frac{\sin(2\pi)}{2} - \frac{\sin(0)}{2} = 0 - 0 = 0$$

Adding the results:

$$A = \pi + 0 = \pi$$

The area enclosed between the polar curves $r = 2$ and $r = 2\sin(\theta)$ is:

$$A = \pi \quad \text{square units}$$

7.7 Sketch the graph of the polar equation $r = 1 + \cos(\theta)$.

To sketch the graph of the polar equation $r = 1 + \cos(\theta)$, we will follow a systematic approach:

• 1. Identify the Type of Curve

The given equation is of the form $r = a + b\cos(\theta)$, which represents a conic section known as a **limacon**. Depending on the values of a and b, the limacon can have different shapes:

- If $a > b$, the limacon has no inner loop.
- If $a = b$, the limacon has a cardioid shape.
- If $a < b$, the limacon has an inner loop.

In our case, $a = 1$ and $b = 1$, so the equation represents a **cardioid**.

• 2. Plot Key Points

To accurately sketch the cardioid, we will calculate the values of r for key angles θ.

θ (radians)	$r = 1 + \cos(\theta)$
0	$1 + \cos(0) = 1 + 1 = 2$
$\frac{\pi}{6}$	$1 + \cos\left(\frac{\pi}{6}\right) = 1 + \frac{\sqrt{3}}{2} \approx 1.866$
$\frac{\pi}{3}$	$1 + \cos\left(\frac{\pi}{3}\right) = 1 + \frac{1}{2} = 1.5$
$\frac{\pi}{2}$	$1 + \cos\left(\frac{\pi}{2}\right) = 1 + 0 = 1$
π	$1 + \cos(\pi) = 1 - 1 = 0$
$\frac{3\pi}{2}$	$1 + \cos\left(\frac{3\pi}{2}\right) = 1 + 0 = 1$
2π	$1 + \cos(2\pi) = 1 + 1 = 2$

• 3. Symmetry Considerations

The equation $r = 1 + \cos(\theta)$ has symmetry about the polar axis (the horizontal axis in polar coordinates) because it depends on $\cos(\theta)$, which is an even function. Therefore, the graph will be symmetric with respect to the polar axis.

• 4. Plotting the Points and Drawing the Curve

Using the key points calculated, we can plot them in polar coordinates:

• 5. Analyzing the Graph

From the plotted points and the symmetry:

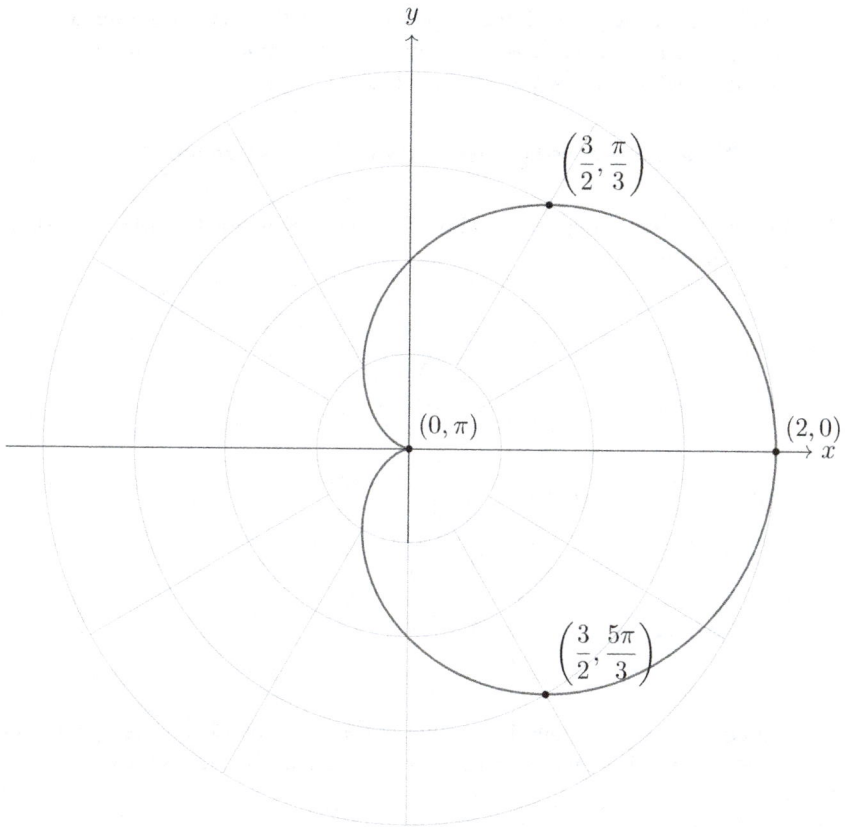

Figure 7.1: Graph of the polar equation $r = 1 + \cos(\theta)$.

- At $\theta = 0$, $r = 2$ (farthest point on the right).

- At $\theta = \pi$, $r = 0$ (polar origin).

- The graph forms a heart-shaped curve known as a cardioid.

- The maximum value of r is 2, and the minimum is 0.

- The symmetry about the polar axis simplifies the sketch, as we only need to plot one side and reflect it.

- **6. Conclusion**

The polar equation $r = 1 + \cos(\theta)$ represents a cardioid. By plotting key points, considering symmetry, and analyzing the behavior of r with respect to

141

θ, we can accurately sketch the graph of the equation. The resulting graph is a smooth, heart-shaped curve symmetric about the polar axis, with its cusp at the origin and its farthest point at $(2,0)$.

- **7. Verification using Cartesian Coordinates (Optional)**

For further verification, we can convert the polar equation to Cartesian coordinates using the relations:

$$x = r\cos(\theta)\ y \quad = r\sin(\theta)$$
$$r = \sqrt{x^2 + y^2}$$
$$\cos(\theta) = \frac{x}{r}$$

Substituting $r = 1 + \cos(\theta)$:

$$r = 1 + \frac{x}{r}$$
$$r^2 = r + x$$
$$x^2 + y^2 = \sqrt{x^2 + y^2} + x$$

This equation is more complex to solve in Cartesian coordinates, which further illustrates the advantage of using polar coordinates for such curves.

- **8. Summary**

Sketching the graph of $r = 1 + \cos(\theta)$ involves recognizing it as a cardioid, plotting key points, exploiting symmetry, and drawing the characteristic heart-shaped curve. Polar coordinates provide a straightforward and efficient method for representing and analyzing such curves.

7.8 Find the slope of the tangent line to $r = \theta$ at $\theta = \pi$.

To determine the slope of the tangent line to the curve defined by the polar equation $r = \theta$ at $\theta = \pi$, we will follow these steps:

1. Express the Polar Coordinates in Cartesian Form:

In polar coordinates, any point on the plane is represented by (r, θ), where r is the radial distance from the origin, and θ is the angle measured from the positive x-axis. To find the slope of the tangent line, it's beneficial to convert the polar equation to Cartesian coordinates using the following transformations:

$$x = r \cos \theta \quad \text{and} \quad y = r \sin \theta$$

Given $r = \theta$, substituting this into the above equations yields:

$$x = \theta \cos \theta \quad \text{and} \quad y = \theta \sin \theta$$

2. Differentiate x and y with respect to θ:

To find the slope $\frac{dy}{dx}$, we first compute $\frac{dy}{d\theta}$ and $\frac{dx}{d\theta}$:

$$\frac{dy}{d\theta} = \frac{d}{d\theta}(\theta \sin \theta) = \sin \theta + \theta \cos \theta$$

$$\frac{dx}{d\theta} = \frac{d}{d\theta}(\theta \cos \theta) = \cos \theta - \theta \sin \theta$$

3. Use the Chain Rule to Find $\frac{dy}{dx}$:

The slope of the tangent line in Cartesian coordinates is given by:

$$\frac{dy}{dx} = \frac{\frac{dy}{d\theta}}{\frac{dx}{d\theta}}$$

Substituting the derivatives computed above:

$$\frac{dy}{dx} = \frac{\sin \theta + \theta \cos \theta}{\cos \theta - \theta \sin \theta}$$

4. Evaluate $\frac{dy}{dx}$ at $\theta = \pi$:

Substitute $\theta = \pi$ into the expression for $\frac{dy}{dx}$:

$$\sin \pi = 0 \quad \text{and} \quad \cos \pi = -1$$

Plugging these values into the numerator and denominator:

$$\frac{dy}{dx}\bigg|_{\theta=\pi} = \frac{0 + \pi \cdot (-1)}{-1 - \pi \cdot 0} = \frac{-\pi}{-1} = \pi$$

5. Conclusion:

The slope of the tangent line to the curve $r = \theta$ at $\theta = \pi$ is π.

Final Answer: The slope of the tangent line at $\theta = \pi$ is π.

$$\boxed{\pi}$$

143

7.9 Compute the length of the curve $r = e^\theta$ from $\theta = 0$ to $\theta = 1$.

To find the length of the curve defined in polar coordinates by $r = e^\theta$ from $\theta = 0$ to $\theta = 1$, we will use the formula for the arc length of a polar curve.

Arc Length Formula in Polar Coordinates

The general formula for the length L of a curve expressed in polar coordinates $r = f(\theta)$ from $\theta = a$ to $\theta = b$ is given by:

$$L = \int_a^b \sqrt{\left(\frac{dr}{d\theta}\right)^2 + r^2}\, d\theta$$

This formula accounts for both the change in the radius r and the change in the angle θ as we move along the curve.

Applying the Formula to $r = e^\theta$

Given the polar equation:

$$r = e^\theta$$

we need to compute $\frac{dr}{d\theta}$. Differentiating both sides with respect to θ:

$$\frac{dr}{d\theta} = \frac{d}{d\theta}\left(e^\theta\right) = e^\theta$$

Substituting $r = e^\theta$ and $\frac{dr}{d\theta} = e^\theta$ into the arc length formula:

$$L = \int_0^1 \sqrt{\left(e^\theta\right)^2 + \left(e^\theta\right)^2}\, d\theta = \int_0^1 \sqrt{2e^{2\theta}}\, d\theta$$

Simplifying under the square root:

$$\sqrt{2e^{2\theta}} = e^\theta \sqrt{2}$$

Thus, the integral becomes:

$$L = \sqrt{2} \int_0^1 e^\theta\, d\theta$$

144

Evaluating the Integral

The integral of e^θ with respect to θ is straightforward:

$$\int e^\theta \, d\theta = e^\theta + C$$

Applying the limits of integration from 0 to 1:

$$\int_0^1 e^\theta \, d\theta = e^1 - e^0 = e - 1$$

Therefore, the length L of the curve is:

$$L = \sqrt{2}(e - 1)$$

Final Answer

The length of the curve $r = e^\theta$ from $\theta = 0$ to $\theta = 1$ is:

$$L = \sqrt{2}\,(e - 1)$$

Numerically, since $e \approx 2.71828$, the length can be approximated as:

$$L \approx \sqrt{2} \times (2.71828 - 1) \approx \sqrt{2} \times 1.71828 \approx 2.429$$

Thus, the curve length is approximately 2.429 units.

7.10 Analyze the conic section defined by $r = \frac{4}{1+\cos(\theta)}$

To analyze the conic section defined by the polar equation $r = \frac{4}{1+\cos(\theta)}$, we will determine its type, key features, and geometric properties. The analysis involves identifying the eccentricity, directrix, focal properties, and sketching the graph based on the given equation.

Standard Form of a Conic Section in Polar Coordinates

Conic sections in polar coordinates, with one focus at the origin, are generally expressed in the standard form:

$$r = \frac{ed}{1 + e\cos(\theta)}$$

where:

145

- e is the eccentricity of the conic section.

- d is the distance from the directrix to the pole (origin).

The eccentricity e determines the type of conic:

- $0 \le e < 1$: Ellipse

- $e = 1$: Parabola

- $e > 1$: Hyperbola

Identifying the Eccentricity and Directrix

Comparing the given equation $r = \frac{4}{1+\cos(\theta)}$ with the standard form $r = \frac{ed}{1+e\cos(\theta)}$, we can equate the numerators:

$$ed = 4$$

The denominator suggests the angle term is $\cos(\theta)$, indicating that the conic is oriented such that the axis is along the polar axis (the positive x-axis).

To find e and d, we observe that e is implicitly present in the denominator. Since the denominator is $1 + \cos(\theta)$, this indicates that:

$$e = 1$$

Substituting $e = 1$ into the equation $ed = 4$:

$$1 \cdot d = 4 \quad \Rightarrow \quad d = 4$$

Determining the Type of Conic Section

Given that $e = 1$, the conic section is a **parabola**. This is because the eccentricity $e = 1$ corresponds to a parabolic shape in conic sections.

Identifying Key Features of the Parabola

For a parabola in polar coordinates with the focus at the origin, the standard equation is:

$$r = \frac{ed}{1 + e\cos(\theta)}$$

Given $e = 1$ and $d = 4$, the equation simplifies to:

146

$$r = \frac{4}{1 + \cos(\theta)}$$

Key features include:

- **Focus**: Located at the origin.

- **Directrix**: The directrix is a line perpendicular to the polar axis. For a parabola defined by $r = \frac{ed}{1+e\cos(\theta)}$, the distance from the directrix to the pole is d. Given $d = 4$, the directrix is the vertical line $\theta = \pi$ (or $x = -4$ in Cartesian coordinates).

- **Axis of Symmetry**: Along the polar axis ($\theta = 0$).

- **Vertex**: The point on the parabola closest to the focus (origin). Substituting $\theta = 0$ into the equation:

$$r = \frac{4}{1 + \cos(0)} = \frac{4}{1 + 1} = 2$$

Therefore, the vertex is at $(r, \theta) = (2, 0)$.

Conversion to Cartesian Coordinates

To gain a deeper understanding, let's convert the polar equation to Cartesian coordinates.

Recall the relationships:

$$r = \sqrt{x^2 + y^2}, \quad \cos(\theta) = \frac{x}{r}$$

Substituting these into the given equation:

$$\sqrt{x^2 + y^2} = \frac{4}{1 + \frac{x}{\sqrt{x^2+y^2}}}$$

Multiply both sides by $1 + \frac{x}{\sqrt{x^2+y^2}}$:

$$\sqrt{x^2 + y^2} \left(1 + \frac{x}{\sqrt{x^2 + y^2}}\right) = 4$$

Simplify:

147

$$\sqrt{x^2 + y^2} + x = 4$$

Isolate $\sqrt{x^2 + y^2}$:

$$\sqrt{x^2 + y^2} = 4 - x$$

Square both sides to eliminate the square root:

$$x^2 + y^2 = (4 - x)^2$$

Expand the right side:

$$x^2 + y^2 = 16 - 8x + x^2$$

Subtract x^2 from both sides:

$$y^2 = 16 - 8x$$

Rearrange to standard Cartesian form:

$$y^2 = -8x + 16 \quad \Rightarrow \quad y^2 = -8(x - 2)$$

This is the equation of a parabola in Cartesian coordinates with:

- **Vertex**: $(2, 0)$

- **Axis of Symmetry**: Along the x-axis

- **Opening Direction**: Since the coefficient of x is negative, the parabola opens to the left.

- **Focus**: At the origin $(0, 0)$

- **Directrix**: The line $x = 4$

Graphical Representation

The parabola described by $r = \frac{4}{1 + \cos(\theta)}$ has the following characteristics:

- It opens to the left, towards negative x-values.

- The vertex is at $(2, 0)$, the closest point to the focus at the origin.

148

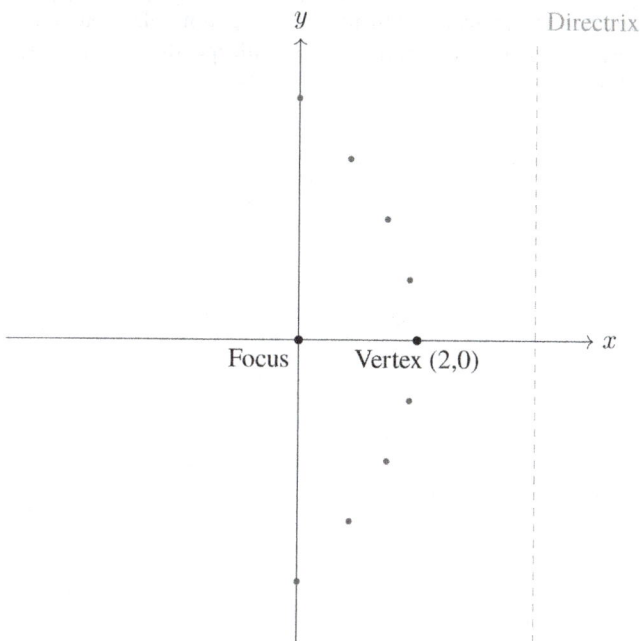

Figure 7.2: Graph of the parabola $r = \frac{4}{1+\cos(\theta)}$

- The directrix is the vertical line $x = 4$, which is the line perpendicular to the axis of symmetry and located at a distance $d = 4$ from the focus.

The polar equation $r = \frac{4}{1+\cos(\theta)}$ represents a parabola with the following properties:

- **Type**: Parabola

- **Eccentricity**: $e = 1$

- **Focus**: Origin $(0,0)$

- **Directrix**: Vertical line $x = 4$

- **Vertex**: $(2,0)$

- **Axis of Symmetry**: x-axis

- **Direction of Opening**: Leftward

149

This comprehensive analysis confirms that the given polar equation defines a parabola oriented along the negative x-axis with specific geometric features as detailed above.

Chapter 8

Multivariable Calculus

This chapter delves into the complexities of multivariable calculus, offering problems that cover differentiation and integration of functions of several variables. Topics include partial derivatives, gradients, multiple integrals, and vector calculus. These problems are designed to enhance understanding of how calculus extends to functions in higher dimensions. Comprehensive solutions provide step-by-step insights, aiding in the development of analytical skills crucial for tackling the multifaceted nature of multivariable calculus.

8.1 Find $\frac{\partial f}{\partial x}$ for $f(x, y) = x^2 y + y^3$.

To find the partial derivative of the function $f(x, y) = x^2 y + y^3$ with respect to x, denoted as $\frac{\partial f}{\partial x}$, we treat y as a constant and differentiate f with respect to x.

Step 1: Understand the Function

The given function is:

$$f(x, y) = x^2 y + y^3$$

Here, y^3 is a term that does not depend on x since y is treated as a constant when taking the partial derivative with respect to x.

Step 2: Apply the Partial Derivative Operator

The partial derivative of f with respect to x is:

$$\frac{\partial f}{\partial x} = \frac{\partial}{\partial x}(x^2 y + y^3)$$

Step 3: Differentiate Term by Term

Differentiate each term of $f(x, y)$ with respect to x:

1. **First Term: $x^2 y$**

$$\frac{\partial}{\partial x}(x^2 y) = y \cdot \frac{\partial}{\partial x}(x^2) = y \cdot 2x = 2xy$$

Since y is treated as a constant, it remains as a multiplier.

2. **Second Term: y^3**

$$\frac{\partial}{\partial x}(y^3) = 0$$

The derivative of a constant with respect to x is zero.

Step 4: Combine the Results

Adding the derivatives of each term:

$$\frac{\partial f}{\partial x} = 2xy + 0 = 2xy$$

The partial derivative of $f(x, y) = x^2 y + y^3$ with respect to x is:

$$\frac{\partial f}{\partial x} = 2xy$$

Verification

To verify, consider specific values of x and y. Let $x = 3$ and $y = 2$:

$$\frac{\partial f}{\partial x} = 2(3)(2) = 12$$

Calculating the derivative manually:

$$f(3, 2) = (3)^2(2) + (2)^3 = 9 \cdot 2 + 8 = 18 + 8 = 26$$

152

A small change in x, say $\Delta x = 0.1$, gives:

$$f(3.1, 2) = (3.1)^2(2) + 8 = 9.61 \cdot 2 + 8 = 19.22 + 8 = 27.22$$

Approximate derivative:

$$\frac{\Delta f}{\Delta x} \approx \frac{27.22 - 26}{0.1} = 12.2$$

This is close to the calculated partial derivative of 12, confirming the correctness of the result.

Final Answer

$$\boxed{\frac{\partial f}{\partial x} = 2xy}$$

8.2 Compute the gradient vector of $f(x, y, z) = xyz$.

The gradient vector of a scalar function $f(x, y, z)$ in three-dimensional space is a vector that consists of all its first-order partial derivatives. It points in the direction of the greatest rate of increase of the function and its magnitude represents the rate of increase in that direction. Mathematically, the gradient of f is denoted by ∇f and is defined as:

$$\nabla f = \left(\frac{\partial f}{\partial x}, \frac{\partial f}{\partial y}, \frac{\partial f}{\partial z} \right)$$

Given the function:

$$f(x, y, z) = xyz$$

We will compute each of its partial derivatives with respect to x, y, and z respectively.

- **Partial Derivative with respect to x**

  To find $\frac{\partial f}{\partial x}$, we treat y and z as constants and differentiate f with respect to x:

  $$\frac{\partial f}{\partial x} = \frac{\partial}{\partial x}(xyz) = yz$$

153

- **Partial Derivative with respect to** y

  Similarly, to find $\frac{\partial f}{\partial y}$, we treat x and z as constants and differentiate f with respect to y:

$$\frac{\partial f}{\partial y} = \frac{\partial}{\partial y}(xyz) = xz$$

- **Partial Derivative with respect to** z

  Finally, to find $\frac{\partial f}{\partial z}$, we treat x and y as constants and differentiate f with respect to z:

$$\frac{\partial f}{\partial z} = \frac{\partial}{\partial z}(xyz) = xy$$

Gradient Vector

Having computed all the necessary partial derivatives, we can now assemble the gradient vector:

$$\nabla f = \left(\frac{\partial f}{\partial x}, \frac{\partial f}{\partial y}, \frac{\partial f}{\partial z} \right) = (yz,\ xz,\ xy)$$

Therefore, the gradient vector of the function $f(x, y, z) = xyz$ is:

$$\nabla f = (yz,\ xz,\ xy)$$

This vector indicates the direction and rate of the steepest ascent of the function f at any point (x, y, z) in its domain.

8.3 Determine the equation of the tangent plane to $z = x^2 + y^2$ at (1,1,2).

To determine the equation of the tangent plane to the surface $z = x^2 + y^2$ at the point $(1, 1, 2)$, we follow these steps:

- **Step 1: Verify the Point Lies on the Surface** First, confirm that the point $(1, 1, 2)$ lies on the surface $z = x^2 + y^2$.

$$z = x^2 + y^2 \qquad z = (1)^2 + (1)^2 = 1 + 1 = 2$$

Since $z = 2$ matches the given point, $(1, 1, 2)$ indeed lies on the surface.

- **Step 2: Compute the Partial Derivatives** The general equation of a tangent plane to a surface $z = f(x, y)$ at a point (x_0, y_0, z_0) is given by:

$$z - z_0 = f_x(x_0, y_0)(x - x_0) + f_y(x_0, y_0)(y - y_0)$$

where f_x and f_y are the partial derivatives of f with respect to x and y, respectively.

Compute the partial derivatives of $f(x, y) = x^2 + y^2$:

$$f_x = \frac{\partial f}{\partial x} = 2x$$

$$f_y = \frac{\partial f}{\partial y} = 2y$$

- **Step 3: Evaluate the Partial Derivatives at the Given Point** Evaluate f_x and f_y at the point $(1, 1)$:

$$f_x(1, 1) = 2(1) = 2$$
$$f_y(1, 1) = 2(1) = 2$$

- **Step 4: Write the Equation of the Tangent Plane** Substitute $x_0 = 1$, $y_0 = 1$, $z_0 = 2$, $f_x(1, 1) = 2$, and $f_y(1, 1) = 2$ into the general equation of the tangent plane:

$$z - 2 = 2(x - 1) + 2(y - 1)$$

- **Step 5: Simplify the Equation** Expand and simplify the equation:

$$z - 2 = 2x - 2 + 2y - 2z - 2 = 2x + 2y - 4z = 2x + 2y - 2$$

The equation of the tangent plane to the surface $z = x^2 + y^2$ at the point $(1, 1, 2)$ is:

$$z = 2x + 2y - 2$$

- **Alternate Form** Alternatively, the equation can be written in the standard form:

$$2x + 2y - z = 2$$

Both forms represent the same plane in three-dimensional space.

- **Verification** To verify, substitute $x = 1$ and $y = 1$ into the equation:

$$z = 2(1) + 2(1) - 2 = 2 + 2 - 2 = 2$$

Which matches the given point $(1, 1, 2)$, confirming the correctness of the tangent plane equation.

8.4 Apply the Chain Rule to $z = x^2 + y^2$ where $x = r\cos(\theta), y = r\sin(\theta)$

In this problem, we aim to apply the chain rule to the function $z = x^2 + y^2$ with the transformations $x = r\cos(\theta)$ and $y = r\sin(\theta)$. This will allow us to express the partial derivatives of z with respect to r and θ.

- First, substitute x and y in terms of r and θ into the equation for z:

$$z = x^2 + y^2 = (r\cos(\theta))^2 + (r\sin(\theta))^2 = r^2\cos^2(\theta) + r^2\sin^2(\theta)$$

- Factor out r^2:
$$z = r^2(\cos^2(\theta) + \sin^2(\theta))$$

- Using the Pythagorean identity $\cos^2(\theta) + \sin^2(\theta) = 1$:

$$z = r^2 \times 1 = r^2$$

Thus, z simplifies to $z = r^2$. However, to demonstrate the application of the chain rule explicitly, we will proceed by computing the partial derivatives directly.

- The chain rule for multivariable functions allows us to compute the derivatives of z with respect to r and θ by considering the intermediate variables x and y.

- Partial Derivative of z with Respect to r:

$$\frac{\partial z}{\partial r} = \frac{\partial z}{\partial x} \cdot \frac{\partial x}{\partial r} + \frac{\partial z}{\partial y} \cdot \frac{\partial y}{\partial r}$$

Compute each component:

$$\frac{\partial z}{\partial x} = 2x$$

156

$$\frac{\partial z}{\partial y} = 2y$$

$$\frac{\partial x}{\partial r} = \cos(\theta)$$

$$\frac{\partial y}{\partial r} = \sin(\theta)$$

Substitute these into the chain rule expression:

$$\frac{\partial z}{\partial r} = 2x \cdot \cos(\theta) + 2y \cdot \sin(\theta)$$

Substitute $x = r\cos(\theta)$ and $y = r\sin(\theta)$:

$$\frac{\partial z}{\partial r} = 2(r\cos(\theta)) \cdot \cos(\theta) + 2(r\sin(\theta)) \cdot \sin(\theta)$$

$$\frac{\partial z}{\partial r} = 2r\cos^2(\theta) + 2r\sin^2(\theta)$$

$$\frac{\partial z}{\partial r} = 2r(\cos^2(\theta) + \sin^2(\theta)) = 2r \times 1 = 2r$$

- Partial Derivative of z with Respect to θ:

$$\frac{\partial z}{\partial \theta} = \frac{\partial z}{\partial x} \cdot \frac{\partial x}{\partial \theta} + \frac{\partial z}{\partial y} \cdot \frac{\partial y}{\partial \theta}$$

Compute each component:

$$\frac{\partial x}{\partial \theta} = -r\sin(\theta)$$

$$\frac{\partial y}{\partial \theta} = r\cos(\theta)$$

Substitute these into the chain rule expression:

$$\frac{\partial z}{\partial \theta} = 2x \cdot (-r\sin(\theta)) + 2y \cdot (r\cos(\theta))$$

Substitute $x = r\cos(\theta)$ and $y = r\sin(\theta)$:

$$\frac{\partial z}{\partial \theta} = 2(r\cos(\theta)) \cdot (-r\sin(\theta)) + 2(r\sin(\theta)) \cdot (r\cos(\theta))$$

$$\frac{\partial z}{\partial \theta} = -2r^2\cos(\theta)\sin(\theta) + 2r^2\sin(\theta)\cos(\theta)$$

$$\frac{\partial z}{\partial \theta} = 0$$

By applying the chain rule, we have determined the partial derivatives of z with respect to r and θ:

$$\frac{\partial z}{\partial r} = 2r \quad \text{and} \quad \frac{\partial z}{\partial \theta} = 0$$

This result is consistent with the earlier simplification $z = r^2$, where it is evident that z depends solely on r and not on θ.

8.5 Calculate the directional derivative of $f(x, y) = xe^y$ in the direction of $\vec{v} = \langle 1, 1 \rangle$.

To find the directional derivative of the function $f(x, y) = xe^y$ in the direction of the vector $\vec{v} = \langle 1, 1 \rangle$, we follow these steps:

- **Step 1: Compute the Gradient of f**

  The gradient vector of a function $f(x, y)$ is given by:

  $$\nabla f = \left\langle \frac{\partial f}{\partial x}, \frac{\partial f}{\partial y} \right\rangle$$

  First, compute the partial derivatives of f with respect to x and y:

  $$\frac{\partial f}{\partial x} = \frac{\partial}{\partial x}(xe^y) = e^y$$

  $$\frac{\partial f}{\partial y} = \frac{\partial}{\partial y}(xe^y) = xe^y$$

  Therefore, the gradient vector is:

  $$\nabla f = \langle e^y, xe^y \rangle$$

- **Step 2: Normalize the Direction Vector $\vec{v}$**

  The directional derivative requires a unit vector in the direction of $\vec{v}$. Given $\vec{v} = \langle 1, 1 \rangle$, the magnitude of $\vec{v}$ is:

  $$\|\vec{v}\| = \sqrt{1^2 + 1^2} = \sqrt{2}$$

  Thus, the unit vector $\vec{u}$ in the direction of $\vec{v}$ is:

  $$\vec{u} = \frac{\vec{v}}{\|\vec{v}\|} = \left\langle \frac{1}{\sqrt{2}}, \frac{1}{\sqrt{2}} \right\rangle$$

- **Step 3: Compute the Directional Derivative**

  The directional derivative of f at a point (x, y) in the direction of $\vec{u}$ is given by the dot product of the gradient vector and the unit direction vector:

  $$D_{\vec{u}} f = \nabla f \cdot \vec{u}$$

  Substituting the expressions for ∇f and $\vec{u}$:

  $$D_{\vec{u}} f = \langle e^y, xe^y \rangle \cdot \left\langle \frac{1}{\sqrt{2}}, \frac{1}{\sqrt{2}} \right\rangle$$

  Compute the dot product:

  $$D_{\vec{u}} f = e^y \left(\frac{1}{\sqrt{2}} \right) + xe^y \left(\frac{1}{\sqrt{2}} \right) = \frac{e^y}{\sqrt{2}} + \frac{xe^y}{\sqrt{2}}$$

  Factor out $\frac{e^y}{\sqrt{2}}$:

  $$D_{\vec{u}} f = \frac{e^y}{\sqrt{2}} (1 + x)$$

The directional derivative of $f(x, y) = xe^y$ in the direction of $\vec{v} = \langle 1, 1 \rangle$ is:

$$D_{\vec{u}} f = \frac{e^y}{\sqrt{2}} (1 + x)$$

This represents the rate at which the function f changes at any point (x, y) in the specified direction.

8.6 Find and classify the critical points of $f(x, y) = x^2 + y^2 - 4x - 6y$.

To find and classify the critical points of the function $f(x, y) = x^2 + y^2 - 4x - 6y$, we will follow these steps:

1. **Find the critical points by setting the first partial derivatives to zero.**

2. **Classify each critical point using the second derivative test.**

Finding Critical Points

First, we find the critical points by computing the first partial derivatives of f with respect to x and y, and then setting them equal to zero.

Step 1: Compute the first partial derivatives.

$$f_x = \frac{\partial f}{\partial x} = 2x - 4$$

$$f_y = \frac{\partial f}{\partial y} = 2y - 6$$

Step 2: Set the partial derivatives equal to zero and solve for x and y.

$$2x - 4 = 0 \quad \Rightarrow \quad x = 2$$
$$2y - 6 = 0 \quad \Rightarrow \quad y = 3$$

Thus, the function $f(x, y)$ has a single critical point at $(2, 3)$.

Classifying Critical Points

To classify the critical point, we use the second derivative test. This involves computing the second partial derivatives and evaluating the discriminant D at the critical point.

Step 1: Compute the second partial derivatives.

$$f_{xx} = \frac{\partial^2 f}{\partial x^2} = 2$$

$$f_{yy} = \frac{\partial^2 f}{\partial y^2} = 2$$

$$f_{xy} = \frac{\partial^2 f}{\partial x \partial y} = 0$$

Step 2: Compute the discriminant D.

$$D = f_{xx}f_{yy} - (f_{xy})^2$$
$$D = (2)(2) - (0)^2 = 4$$

Step 3: Analyze the discriminant D and f_{xx} to classify the critical point.

- If $D > 0$ and $f_{xx} > 0$, then f has a local minimum at the critical point.

- If $D > 0$ and $f_{xx} < 0$, then f has a local maximum at the critical point.

- If $D < 0$, then f has a saddle point at the critical point.

- If $D = 0$, the test is inconclusive.

In our case:

$$D = 4 > 0 \quad \text{and} \quad f_{xx} = 2 > 0$$

Therefore, the function $f(x, y)$ has a **local minimum** at the critical point $(2, 3)$. The function $f(x, y) = x^2 + y^2 - 4x - 6y$ has a single critical point at $(2, 3)$, which is a **local minimum**.

Verification by Completing the Square

For further verification, we can complete the square for both x and y:

$$f(x, y) = x^2 - 4x + y^2 - 6y$$
$$= (x^2 - 4x + 4) - 4 + (y^2 - 6y + 9) - 9$$
$$= (x - 2)^2 + (y - 3)^2 - 13$$

Since $(x - 2)^2$ and $(y - 3)^2$ are both squares of real numbers and hence non-negative, the minimum value of $f(x, y)$ occurs when both squares are zero, i.e., at $(2, 3)$. This confirms that $f(x, y)$ attains its minimum value of -13 at the critical point $(2, 3)$.

8.7 Use Lagrange multipliers to find the extrema of $f(x, y) = xy$ subject to $x^2 + y^2 = 1$.

To find the extrema of the function $f(x, y) = xy$ subject to the constraint $x^2 + y^2 = 1$, we will employ the method of Lagrange multipliers. This method is particularly useful for finding the local maxima and minima of a function subject to equality constraints.

Setting Up the Lagrangian

First, define the constraint equation and the function to be optimized:

$$g(x, y) = x^2 + y^2 - 1 = 0$$

The Lagrangian function $\mathcal{L}$ is constructed by combining the objective function $f(x, y)$ and the constraint $g(x, y)$ multiplied by a Lagrange multiplier λ:

$$\mathcal{L}(x, y, \lambda) = f(x, y) - \lambda g(x, y) = xy - \lambda(x^2 + y^2 - 1)$$

Finding the Critical Points

To find the extrema, we need to find the critical points of the Lagrangian by taking the partial derivatives with respect to x, y, and λ, and setting them equal to zero.

$$\frac{\partial \mathcal{L}}{\partial x} = y - \lambda \cdot 2x = 0 \quad \Rightarrow \quad y = 2\lambda x$$

$$\frac{\partial \mathcal{L}}{\partial y} = x - \lambda \cdot 2y = 0 \quad \Rightarrow \quad x = 2\lambda y$$

$$\frac{\partial \mathcal{L}}{\partial \lambda} = -(x^2 + y^2 - 1) = 0 \quad \Rightarrow \quad x^2 + y^2 = 1$$

Solving the System of Equations

From the first two equations, we have:

$$y = 2\lambda x \quad \text{and} \quad x = 2\lambda y$$

Substituting $y = 2\lambda x$ into the second equation:

$$x = 2\lambda(2\lambda x) = 4\lambda^2 x$$

Assuming $x \neq 0$ (we will consider the case $x = 0$ separately), we can divide both sides by x:

$$1 = 4\lambda^2 \quad \Rightarrow \quad \lambda^2 = \frac{1}{4} \quad \Rightarrow \quad \lambda = \pm\frac{1}{2}$$

Case 1: $\lambda = \frac{1}{2}$

Substituting $\lambda = \frac{1}{2}$ into $y = 2\lambda x$:

$$y = 2\left(\frac{1}{2}\right)x = x$$

Using the constraint $x^2 + y^2 = 1$ and $y = x$:

$$x^2 + x^2 = 1 \quad \Rightarrow \quad 2x^2 = 1 \quad \Rightarrow \quad x^2 = \frac{1}{2} \quad \Rightarrow \quad x = \pm\frac{\sqrt{2}}{2}$$

Therefore, the points are:

$$\left(\frac{\sqrt{2}}{2}, \frac{\sqrt{2}}{2}\right) \quad \text{and} \quad \left(-\frac{\sqrt{2}}{2}, -\frac{\sqrt{2}}{2}\right)$$

Case 2: $\lambda = -\frac{1}{2}$

Substituting $\lambda = -\frac{1}{2}$ into $y = 2\lambda x$:

$$y = 2\left(-\frac{1}{2}\right)x = -x$$

Using the constraint $x^2 + y^2 = 1$ and $y = -x$:

$$x^2 + (-x)^2 = 1 \quad \Rightarrow \quad 2x^2 = 1 \quad \Rightarrow \quad x^2 = \frac{1}{2} \quad \Rightarrow \quad x = \pm\frac{\sqrt{2}}{2}$$

Therefore, the points are:

$$\left(\frac{\sqrt{2}}{2}, -\frac{\sqrt{2}}{2}\right) \quad \text{and} \quad \left(-\frac{\sqrt{2}}{2}, \frac{\sqrt{2}}{2}\right)$$

Case 3: $x = 0$

If $x = 0$, substitute into the constraint $x^2 + y^2 = 1$:

$$0 + y^2 = 1 \quad \Rightarrow \quad y^2 = 1 \quad \Rightarrow \quad y = \pm 1$$

Similarly, if $y = 0$, substitute into the constraint to find $x = \pm 1$. However, substituting $x = 0$ or $y = 0$ into the objective function $f(x, y) = xy$ yields zero, which may correspond to saddle points rather than extrema.

163

Evaluating the Objective Function at the Critical Points

Now, evaluate $f(x, y) = xy$ at each of the critical points found.

- At $\left(\frac{\sqrt{2}}{2}, \frac{\sqrt{2}}{2} \right)$:

$$f\left(\frac{\sqrt{2}}{2}, \frac{\sqrt{2}}{2} \right) = \frac{\sqrt{2}}{2} \cdot \frac{\sqrt{2}}{2} = \frac{2}{4} = \frac{1}{2}$$

- At $\left(-\frac{\sqrt{2}}{2}, -\frac{\sqrt{2}}{2} \right)$:

$$f\left(-\frac{\sqrt{2}}{2}, -\frac{\sqrt{2}}{2} \right) = \left(-\frac{\sqrt{2}}{2} \right) \cdot \left(-\frac{\sqrt{2}}{2} \right) = \frac{2}{4} = \frac{1}{2}$$

- At $\left(\frac{\sqrt{2}}{2}, -\frac{\sqrt{2}}{2} \right)$:

$$f\left(\frac{\sqrt{2}}{2}, -\frac{\sqrt{2}}{2} \right) = \frac{\sqrt{2}}{2} \cdot \left(-\frac{\sqrt{2}}{2} \right) = -\frac{2}{4} = -\frac{1}{2}$$

- At $\left(-\frac{\sqrt{2}}{2}, \frac{\sqrt{2}}{2} \right)$:

$$f\left(-\frac{\sqrt{2}}{2}, \frac{\sqrt{2}}{2} \right) = \left(-\frac{\sqrt{2}}{2} \right) \cdot \frac{\sqrt{2}}{2} = -\frac{2}{4} = -\frac{1}{2}$$

The function $f(x, y) = xy$ subject to the constraint $x^2 + y^2 = 1$ attains its maximum value of $\frac{1}{2}$ at the points $\left(\frac{\sqrt{2}}{2}, \frac{\sqrt{2}}{2} \right)$ and $\left(-\frac{\sqrt{2}}{2}, -\frac{\sqrt{2}}{2} \right)$, and its minimum value of $-\frac{1}{2}$ at the points $\left(\frac{\sqrt{2}}{2}, -\frac{\sqrt{2}}{2} \right)$ and $\left(-\frac{\sqrt{2}}{2}, \frac{\sqrt{2}}{2} \right)$.

These results indicate that the extrema of the function $f(x, y) = xy$ on the unit circle occur at the points where $x = \pm y$, which are the lines $y = x$ and $y = -x$ intersecting the unit circle.

$\square$

8.8 Evaluate $\int_0^1 \int_0^1 xy \, dx \, dy$.

Proof. We aim to evaluate the double integral

$$\int_0^1 \int_0^1 xy \, dx \, dy.$$

This integral represents the volume under the surface $z = xy$ over the unit square in the xy-plane.

Step 1: Integrate with respect to x

First, we perform the inner integral with respect to x, treating y as a constant:

$$\int_0^1 xy \, dx.$$

Since y is constant with respect to x, we can factor it out of the integral:

$$y \int_0^1 x \, dx.$$

Now, compute the integral of x with respect to x:

$$\int x \, dx = \frac{x^2}{2} + C.$$

Evaluating from 0 to 1:

$$y \left[\frac{x^2}{2} \right]_0^1 = y \left(\frac{1^2}{2} - \frac{0^2}{2} \right) = y \left(\frac{1}{2} \right) = \frac{y}{2}.$$

Step 2: Integrate with respect to y

Next, we perform the outer integral with respect to y:

$$\int_0^1 \frac{y}{2} \, dy.$$

Factor out the constant $\frac{1}{2}$:

$$\frac{1}{2} \int_0^1 y \, dy.$$

Compute the integral of y with respect to y:

$$\int y \, dy = \frac{y^2}{2} + C.$$

Evaluating from 0 to 1:

$$\frac{1}{2} \left[\frac{y^2}{2} \right]_0^1 = \frac{1}{2} \left(\frac{1^2}{2} - \frac{0^2}{2} \right) = \frac{1}{2} \left(\frac{1}{2} \right) = \frac{1}{4}.$$

Thus, the value of the double integral is

$$\boxed{\frac{1}{4}}.$$

$\square$

8.9 Compute $\int_0^1 \int_0^{1-x} (x+y)\, dy\, dx.$

To compute the double integral

$$\int_0^1 \int_0^{1-x} (x+y)\, dy\, dx,$$

we will evaluate the integral step by step, starting with the inner integral.

• **Step 1: Evaluate the Inner Integral**

First, fix x and integrate with respect to y:

$$\int_0^{1-x} (x+y)\, dy.$$

This integral can be separated into two simpler integrals:

$$\int_0^{1-x} x\, dy + \int_0^{1-x} y\, dy.$$

Evaluate each integral separately.

$$\int_0^{1-x} x\, dy = x \int_0^{1-x} dy = x\, [y]_0^{1-x} = x(1-x) - x(0) = x(1-x),$$

$$\int_0^{1-x} y\, dy = \left[\frac{y^2}{2}\right]_0^{1-x} = \frac{(1-x)^2}{2} - \frac{0^2}{2} = \frac{(1-x)^2}{2}.$$

Adding these results together:

$$\int_0^{1-x} (x+y)\, dy = x(1-x) + \frac{(1-x)^2}{2}.$$

• **Step 2: Simplify the Expression**

Combine the terms:

$$x(1-x) + \frac{(1-x)^2}{2} = x - x^2 + \frac{1 - 2x + x^2}{2}.$$

166

Distribute and combine like terms:

$$x - x^2 + \frac{1}{2} - x + \frac{x^2}{2} = \frac{1}{2} - \frac{x^2}{2}.$$

- **Step 3: Evaluate the Outer Integral**

Now, integrate the simplified expression with respect to x:

$$\int_0^1 \left(\frac{1}{2} - \frac{x^2}{2} \right) dx.$$

Separate the integral:

$$\frac{1}{2} \int_0^1 dx - \frac{1}{2} \int_0^1 x^2 \, dx.$$

Compute each integral:

$$\frac{1}{2} \int_0^1 dx = \frac{1}{2} [x]_0^1 = \frac{1}{2}(1 - 0) = \frac{1}{2},$$

$$\frac{1}{2} \int_0^1 x^2 \, dx = \frac{1}{2} \left[\frac{x^3}{3} \right]_0^1 = \frac{1}{2} \left(\frac{1}{3} - 0 \right) = \frac{1}{6}.$$

Subtract the second integral from the first:

$$\frac{1}{2} - \frac{1}{6} = \frac{3}{6} - \frac{1}{6} = \frac{2}{6} = \frac{1}{3}.$$

Thus, the value of the double integral is:

$$\int_0^1 \int_0^{1-x} (x+y) \, dy \, dx = \frac{1}{3}.$$

8.10 Change variables to evaluate $\int_0^\infty \int_0^\infty e^{-x^2-y^2} \, dy \, dx$.

To evaluate the double integral

$$\int_0^\infty \int_0^\infty e^{-x^2-y^2} \, dy \, dx,$$

we can utilize a change of variables to polar coordinates. This approach simplifies the integration of functions involving $x^2 + y^2$.

- Recognize the Symmetry and Choose Polar Coordinates

The integrand $e^{-x^2-y^2}$ is radially symmetric, depending only on the distance $r = \sqrt{x^2 + y^2}$. This symmetry suggests that switching to polar coordinates will simplify the integral.

In polar coordinates, the relationships between Cartesian and polar variables are:

$$x = r \cos \theta, \quad y = r \sin \theta,$$

where $r \geq 0$ and $0 \leq \theta \leq \frac{\pi}{2}$ to cover the first quadrant.

The Jacobian determinant for the transformation from Cartesian to polar coordinates is r, so the area element transforms as:

$$dx \, dy = r \, dr \, d\theta.$$

- Rewrite the Integral in Polar Coordinates

Substituting the polar coordinates into the integral, we have:

$$\int_0^\infty \int_0^\infty e^{-x^2-y^2} \, dy \, dx = \int_0^{\frac{\pi}{2}} \int_0^\infty e^{-r^2} r \, dr \, d\theta.$$

- Evaluate the Inner Integral

First, evaluate the integral with respect to r:

$$\int_0^\infty e^{-r^2} r \, dr.$$

Letting $u = r^2$, we have $du = 2r \, dr$, which implies $r \, dr = \frac{1}{2} du$. The limits remain from $u = 0$ to $u = \infty$.

Substituting:

$$\int_0^\infty e^{-r^2} r \, dr = \frac{1}{2} \int_0^\infty e^{-u} \, du = \frac{1}{2} \left[-e^{-u} \right]_0^\infty = \frac{1}{2}.$$

• Evaluate the Outer Integral

Now, integrate with respect to θ:

$$\int_0^{\frac{\pi}{2}} \frac{1}{2} \, d\theta = \frac{1}{2} [\theta]_0^{\frac{\pi}{2}} = \frac{1}{2} \cdot \frac{\pi}{2} = \frac{\pi}{4}.$$

Thus, the value of the double integral is:

$$\int_0^\infty \int_0^\infty e^{-x^2-y^2} \, dy \, dx = \frac{\pi}{4}.$$

Chapter 9

Multiple Integrals and Vector Calculus

This chapter focuses on the advanced topics of multiple integrals and vector calculus, offering problems that address double and triple integrals, line integrals, surface integrals, and theorems such as Green's, Stokes', and the Divergence Theorem. The problems are crafted to build a robust understanding of how multiple integrals and vector fields are used to solve complex problems in higher-dimensional spaces. Detailed solutions provide clarity and facilitate mastery of these critical concepts, essential for advanced study and applications in mathematics and physics.

9.1 Evaluate $\iint_D (x + y) \, dA$ over the region D

To evaluate the double integral

$$\iint_D (x + y) \, dA,$$

we first need to describe the region D and set up the appropriate limits of integration.

- The region D is bounded by the following lines:
    - $x = 0$,
    - $y = 0$,

$- x + y = 1.$

This forms a right triangle in the first quadrant with vertices at $(0, 0)$, $(1, 0)$, and $(0, 1)$.

- We can set up the double integral by integrating y from 0 to $1 - x$ and x from 0 to 1. Thus, the integral becomes:

$$\iint_D (x + y)\, dA = \int_0^1 \int_0^{1-x} (x + y)\, dy\, dx.$$

- First, integrate with respect to y:

$$\int_0^{1-x} (x + y)\, dy = \int_0^{1-x} x\, dy + \int_0^{1-x} y\, dy$$

$$= x\, [y]_0^{1-x} + \left[\frac{y^2}{2}\right]_0^{1-x}$$

$$= x(1 - x) + \frac{(1 - x)^2}{2}$$

$$= x - x^2 + \frac{1 - 2x + x^2}{2}$$

$$= x - x^2 + \frac{1}{2} - x + \frac{x^2}{2}$$

$$= (x - x) + \left(-x^2 + \frac{x^2}{2}\right) + \frac{1}{2}$$

$$= -\frac{x^2}{2} + \frac{1}{2}.$$

Next, integrate the result with respect to x:

$$\int_0^1 \left(-\frac{x^2}{2} + \frac{1}{2}\right) dx = -\frac{1}{2}\int_0^1 x^2\, dx + \frac{1}{2}\int_0^1 dx$$

$$= -\frac{1}{2}\left[\frac{x^3}{3}\right]_0^1 + \frac{1}{2}[x]_0^1$$

$$= -\frac{1}{2}\left(\frac{1}{3} - 0\right) + \frac{1}{2}(1 - 0)$$

$$= -\frac{1}{6} + \frac{1}{2}$$

$$= \frac{1}{3}.$$

Therefore, the value of the double integral is

$$\iint_D (x + y) \, dA = \frac{1}{3}.$$

9.2 Compute $\int_0^{2\pi} \int_0^1 r^3 \, dr \, d\theta$

To compute the double integral

$$\int_0^{2\pi} \int_0^1 r^3 \, dr \, d\theta,$$

we will evaluate the integral step by step, first integrating with respect to r and then with respect to θ.

- **Step 1: Integrate with Respect to r**

Consider the inner integral:

$$\int_0^1 r^3 \, dr.$$

This is a straightforward power function integral. The general form of the integral of r^n with respect to r is:

$$\int r^n \, dr = \frac{r^{n+1}}{n+1} + C,$$

where C is the constant of integration.

Applying this to our integral where $n = 3$:

$$\int r^3 \, dr = \frac{r^4}{4} + C.$$

Now, evaluate the definite integral from $r = 0$ to $r = 1$:

$$\frac{r^4}{4} \bigg|_0^1 = \frac{1^4}{4} - \frac{0^4}{4} = \frac{1}{4} - 0 = \frac{1}{4}.$$

- **Step 2: Integrate with Respect to θ**

173

Now, substitute the result of the inner integral into the outer integral:

$$\int_0^{2\pi} \left(\frac{1}{4}\right) d\theta = \frac{1}{4} \int_0^{2\pi} d\theta.$$

The integral $\int_0^{2\pi} d\theta$ is simply the length of the interval $[0, 2\pi]$, which is 2π. Therefore:

$$\frac{1}{4} \times 2\pi = \frac{2\pi}{4} = \frac{\pi}{2}.$$

The value of the double integral is:

$$\int_0^{2\pi} \int_0^1 r^3 \, dr \, d\theta = \frac{\pi}{2}.$$

9.3 Use Green's Theorem to evaluate $\oint_C (y \, dx - x \, dy)$ where C is the unit circle

To evaluate the line integral

$$\oint_C (y \, dx - x \, dy),$$

where C is the unit circle, we will apply Green's Theorem. Green's Theorem relates a line integral around a simple, closed, positively oriented curve C in the plane to a double integral over the region D enclosed by C.

Statement of Green's Theorem

Green's Theorem states that:

$$\oint_C (P \, dx + Q \, dy) = \iint_D \left(\frac{\partial Q}{\partial x} - \frac{\partial P}{\partial y}\right) dA,$$

where:

- C is a positively oriented, piecewise smooth, simple closed curve in the plane,

- D is the region bounded by C,

- P and Q are functions of (x, y) with continuous partial derivatives on an open region containing D.

Identifying P and Q

In our integral, we have:

$$P = y \quad \text{and} \quad Q = -x.$$

We need to compute the partial derivatives $\frac{\partial Q}{\partial x}$ and $\frac{\partial P}{\partial y}$.

Computing the Partial Derivatives

$$\frac{\partial Q}{\partial x} = \frac{\partial(-x)}{\partial x} = -1,$$

$$\frac{\partial P}{\partial y} = \frac{\partial y}{\partial y} = 1.$$

Setting Up the Double Integral

Substituting P and Q into Green's Theorem:

$$\oint_C (y\,dx - x\,dy) = \iint_D \left(\frac{\partial(-x)}{\partial x} - \frac{\partial y}{\partial y} \right) dA = \iint_D (-1-1)\,dA = \iint_D (-2)\,dA.$$

Evaluating the Double Integral

Since D is the unit circle, its area is:

$$\text{Area of } D = \pi r^2 = \pi \cdot 1^2 = \pi.$$

Thus, the double integral becomes:

$$\iint_D (-2)\,dA = -2 \cdot \text{Area of } D = -2\pi.$$

Therefore, the value of the line integral is:

$$\oint_C (y\,dx - x\,dy) = -2\pi.$$

Verification by Parametrization (Optional)

For verification, consider parametrizing the unit circle C using:

$$x = \cos\theta, \quad y = \sin\theta, \quad \theta \in [0, 2\pi].$$

Then:

$$dx = -\sin\theta\,d\theta, \quad dy = \cos\theta\,d\theta.$$

Substituting into the integral:

175

$$\oint_C (y\,dx - x\,dy) = \int_0^{2\pi} (\sin\theta)(-\sin\theta\,d\theta) - (\cos\theta)(\cos\theta\,d\theta)$$

$$= \int_0^{2\pi} (-\sin^2\theta - \cos^2\theta)\,d\theta.$$

Simplifying:

$$-\int_0^{2\pi} (\sin^2\theta + \cos^2\theta)\,d\theta = -\int_0^{2\pi} 1\,d\theta = -2\pi,$$

which confirms our result obtained via Green's Theorem.

Final Answer

$$\oint_C (y\,dx - x\,dy) = -2\pi.$$

9.4 Calculate the surface integral of $F(x, y, z) = x\hat{i} + y\hat{j} + z\hat{k}$ over the unit sphere

To compute the surface integral of the vector field $F(x, y, z) = x\hat{i} + y\hat{j} + z\hat{k}$ over the unit sphere S, we can employ two primary methods: the direct calculation using parameterization and the application of the Divergence Theorem. Below, both methods are detailed for clarity and completeness.

Method 1: Direct Calculation Using Parameterization

Parameterization of the Unit Sphere

The unit sphere S in $\mathbb{R}^3$ is defined by the equation:

$$x^2 + y^2 + z^2 = 1$$

We can parameterize the unit sphere using spherical coordinates:

$$\begin{cases} x = \sin\phi\cos\theta, \\ y = \sin\phi\sin\theta, \\ z = \cos\phi, \end{cases}$$

where $\phi \in [0, \pi]$ is the polar angle and $\theta \in [0, 2\pi)$ is the azimuthal angle.

Surface Element dS

176

The differential surface element on the sphere can be expressed as:

$$dS = n \, dS$$

where n is the outward unit normal vector to the surface, and dS is the scalar area element.

For the unit sphere, the outward unit normal vector is the position vector itself:

$$n = \frac{r}{|r|} = x\hat{i} + y\hat{j} + z\hat{k}$$

since $|r| = 1$ on the unit sphere.

The scalar area element in spherical coordinates is:

$$dS = \sin\phi \, d\phi \, d\theta$$

Thus, the vector surface element becomes:

$$dS = (x\hat{i} + y\hat{j} + z\hat{k}) \sin\phi \, d\phi \, d\theta$$

Evaluating the Surface Integral

The surface integral of F over S is given by:

$$\iint_S F \cdot dS = \iint_S (x\hat{i} + y\hat{j} + z\hat{k}) \cdot (x\hat{i} + y\hat{j} + z\hat{k}) \sin\phi \, d\phi \, d\theta$$

Simplifying the dot product:

$$F \cdot dS = (x^2 + y^2 + z^2) \sin\phi \, d\phi \, d\theta$$

onumber On the unit sphere, $x^2 + y^2 + z^2 = 1$, so:

$$F \cdot dS = \sin\phi \, d\phi \, d\theta$$

Therefore, the integral becomes:

$$\iint_S F \cdot dS = \int_0^{2\pi} \int_0^{\pi} \sin\phi \, d\phi \, d\theta$$

First, integrate with respect to ϕ:

$$\int_0^{\pi} \sin\phi \, d\phi = [-\cos\phi]_0^{\pi} = -\cos\pi + \cos 0 = 1 + 1 = 2$$

Next, integrate with respect to θ:

$$\int_0^{2\pi} 2 \, d\theta = 2\theta \Big|_0^{2\pi} = 4\pi$$

177

Thus, the surface integral evaluates to:

$$\iint_S \mathbf{F} \cdot d\mathbf{S} = 4\pi$$

Method 2: Application of the Divergence Theorem

Statement of the Divergence Theorem

The Divergence Theorem relates the flux of a vector field through a closed surface to the divergence of the field within the volume bounded by the surface:

$$\iint_S \mathbf{F} \cdot d\mathbf{S} = \iiint_V \nabla \cdot \mathbf{F}\, dV$$

where:

- S is the closed surface bounding the volume V.

- $\mathbf{F}$ is a continuously differentiable vector field.

- $\nabla \cdot \mathbf{F}$ is the divergence of $\mathbf{F}$.

Computing the Divergence of F

Given the vector field:

$$\mathbf{F}(x, y, z) = x\hat{i} + y\hat{j} + z\hat{k}$$

The divergence $\nabla \cdot \mathbf{F}$ is:

$$\nabla \cdot \mathbf{F} = \frac{\partial}{\partial x}(x) + \frac{\partial}{\partial y}(y) + \frac{\partial}{\partial z}(z) = 1 + 1 + 1 = 3$$

Computing the Volume Integral

The volume V bounded by the unit sphere has volume:

$$V = \frac{4}{3}\pi r^3 = \frac{4}{3}\pi(1)^3 = \frac{4}{3}\pi$$

Since $\nabla \cdot \mathbf{F} = 3$ is constant within V, the volume integral simplifies to:

$$\iiint_V \nabla \cdot \mathbf{F}\, dV = 3 \cdot \frac{4}{3}\pi = 4\pi$$

Conclusion via the Divergence Theorem

Applying the Divergence Theorem, we have:

$$\iint_S F \cdot dS = \iiint_V \nabla \cdot F \, dV = 4\pi$$

Thus, the surface integral of F over the unit sphere is:

$$\iint_S F \cdot dS = 4\pi$$

Final Answer

Both methods yield the same result. Therefore, the surface integral of $F(x, y, z) = x\hat{i} + y\hat{j} + z\hat{k}$ over the unit sphere is:

$$\iint_S F \cdot dS = 4\pi$$

9.5 Apply the Divergence Theorem to $F = x\hat{i} + y\hat{j} + z\hat{k}$ in a Sphere of Radius R

The Divergence Theorem, also known as Gauss's Theorem, provides a relationship between the flux of a vector field through a closed surface and the divergence of the field within the volume bounded by that surface. Specifically, the theorem states:

$$\iint_{\partial V} F \cdot n \, dS = \iiint_V \nabla \cdot F \, dV$$

where:

- F is a continuously differentiable vector field,
- V is a compact subset of $\mathbb{R}^3$ with boundary ∂V that is piecewise smooth,
- n is the outward-pointing unit normal vector on ∂V,
- dS is the differential surface element,
- dV is the differential volume element.

In this problem, we are to apply the Divergence Theorem to the vector field $F = x\hat{i} + y\hat{j} + z\hat{k}$ over a sphere of radius R.

179

Step 1: Compute the Divergence of F

First, we need to compute the divergence $\nabla \cdot F$.

Given:
$$F = x\hat{i} + y\hat{j} + z\hat{k}$$

The divergence is:
$$\nabla \cdot F = \frac{\partial F_x}{\partial x} + \frac{\partial F_y}{\partial y} + \frac{\partial F_z}{\partial z}$$

Calculating each term:
$$\frac{\partial F_x}{\partial x} = \frac{\partial}{\partial x}(x) = 1$$

$$\frac{\partial F_y}{\partial y} = \frac{\partial}{\partial y}(y) = 1$$

$$\frac{\partial F_z}{\partial z} = \frac{\partial}{\partial z}(z) = 1$$

Thus:
$$\nabla \cdot F = 1 + 1 + 1 = 3$$

Step 2: Set Up the Volume Integral

According to the Divergence Theorem:

$$\iint_{\partial V} F \cdot n \, dS = \iiint_V \nabla \cdot F \, dV = \iiint_V 3 \, dV$$

Since the sphere has radius R, its volume V is:

$$V = \frac{4}{3}\pi R^3$$

Therefore:
$$\iiint_V 3 \, dV = 3 \times \frac{4}{3}\pi R^3 = 4\pi R^3$$

Step 3: Conclude the Result

By the Divergence Theorem, the flux of F through the surface ∂V of the sphere is equal to the volume integral of the divergence of F over V. Therefore:

$$\iint_{\partial V} F \cdot n \, dS = 4\pi R^3$$

Verification (Optional): Compute the Surface Integral Directly

For completeness, let us verify the result by directly computing the surface integral.

Parametrization of the Sphere

The sphere of radius R can be parametrized in spherical coordinates as:

$$r(\theta, \phi) = R\sin\phi\cos\theta\,\hat{i} + R\sin\phi\sin\theta\,\hat{j} + R\cos\phi\,\hat{k}$$

where $\theta \in [0, 2\pi)$ and $\phi \in [0, \pi]$.

The outward unit normal vector n is given by:

$$n = \frac{r}{|r|} = \sin\phi\cos\theta\,\hat{i} + \sin\phi\sin\theta\,\hat{j} + \cos\phi\,\hat{k}$$

Evaluating $F \cdot n$

Given $F = x\hat{i} + y\hat{j} + z\hat{k}$, and using the parametrization:

$$F = R\sin\phi\cos\theta\,\hat{i} + R\sin\phi\sin\theta\,\hat{j} + R\cos\phi\,\hat{k}$$

Thus:

$$F\cdot n = (R\sin\phi\cos\theta)(\sin\phi\cos\theta)+(R\sin\phi\sin\theta)(\sin\phi\sin\theta)+(R\cos\phi)(\cos\phi)$$

$$= R\sin^2\phi\cos^2\theta + R\sin^2\phi\sin^2\theta + R\cos^2\phi$$
$$= R\sin^2\phi(\cos^2\theta + \sin^2\theta) + R\cos^2\phi$$
$$= R\sin^2\phi(1) + R\cos^2\phi$$
$$= R(\sin^2\phi + \cos^2\phi) = R$$

Computing the Surface Integral

The surface integral becomes:

$$\iint_{\partial V} F\cdot n\,dS = \iint_{\partial V} R\,dS = R\iint_{\partial V} dS$$

The surface area of the sphere is:

$$\iint_{\partial V} dS = 4\pi R^2$$

Therefore:

$$\iint_{\partial V} F\cdot n\,dS = R \times 4\pi R^2 = 4\pi R^3$$

181

This matches the result obtained using the Divergence Theorem, confirming the correctness of our application.

By applying the Divergence Theorem to the vector field $\boldsymbol{F} = x\hat{i} + y\hat{j} + z\hat{k}$ over a sphere of radius R, we have determined that the flux through the surface of the sphere is:

$$\iint_{\partial V} \boldsymbol{F} \cdot \boldsymbol{n}\, dS = 4\pi R^3$$

This result aligns with the direct computation of the surface integral, thereby validating the application of the Divergence Theorem in this context.

9.6 Evaluate the line integral $\int_C (x^2\hat{i} + y^2\hat{j}) \cdot d\vec{r}$ along a given path

To evaluate the line integral

$$\int_C \left(x^2\hat{i} + y^2\hat{j}\right) \cdot d\vec{r},$$

we first need to understand the nature of the path C. Since the path is not explicitly provided, we will consider two common scenarios:

- Path C is a closed curve, allowing us to apply Green's Theorem.

- Path C is parametrized, enabling direct evaluation of the integral.

- For a closed curve C: Using Green's Theorem, the line integral $\int_C (x^2\hat{i} + y^2\hat{j}) \cdot d\vec{r}$ evaluates to zero.

- For a specific parametrized path C: The integral can be computed by substituting the parametric equations into the integral and evaluating accordingly. In the example of a straight line from $(0,0)$ to $(1,1)$, the integral evaluates to $\frac{2}{3}$.

The method of evaluation depends critically on the nature of the path C. For generalized paths, especially closed curves, applying Green's Theorem simplifies the computation significantly.

182

Algorithm 1 Using Green's Theorem (Closed Curve)

Assume that C is a positively oriented, simple, closed curve enclosing a region D in the plane. Green's Theorem relates the line integral around C to a double integral over the region D:

$$\oint_C (P\,dx + Q\,dy) = \iint_D \left(\frac{\partial Q}{\partial x} - \frac{\partial P}{\partial y} \right) dA.$$

In our case, $P = x^2$ and $Q = y^2$. Compute the partial derivatives:

$$\frac{\partial Q}{\partial x} = \frac{\partial}{\partial x}(y^2) = 0,$$

$$\frac{\partial P}{\partial y} = \frac{\partial}{\partial y}(x^2) = 0.$$

Substituting into Green's Theorem:

$$\oint_C \left(x^2\,dx + y^2\,dy \right) = \iint_D (0 - 0)\,dA = 0.$$

Therefore, for any closed curve C, the line integral evaluates to zero.

Final Answer Depending on the path C, the line integral evaluates as follows:

- **Closed Path:** $\int_C \left(x^2\,dx + y^2\,dy \right) = 0.$

- **Specific Path (e.g., straight line from $(0,0)$ to $(1,1)$):** $\int_C \left(x^2\,dx + y^2\,dy \right) = \frac{2}{3}.$

9.7 Determine if the vector field $F = y\hat{i} - x\hat{j}$ is conservative

To determine whether the vector field $\boldsymbol{F} = y\hat{i} - x\hat{j}$ is conservative, we will analyze the properties of conservative vector fields and apply the necessary criteria.

Definition of a Conservative Vector Field

A vector field $\boldsymbol{F} = P(x,y)\hat{i} + Q(x,y)\hat{j} + R(x,y,z)\hat{k}$ is said to be **conservative**

Algorithm 2 Direct Evaluation (Parametrized Path)

Alternatively, suppose that the path C is parametrized by $\vec{r}(t) = \langle x(t), y(t) \rangle$ for $a \leq t \leq b$. The line integral can be evaluated directly as follows:

$$\int_C \left(x^2 \hat{i} + y^2 \hat{j} \right) \cdot d\vec{r} = \int_a^b \left(x^2 \frac{dx}{dt} + y^2 \frac{dy}{dt} \right) dt.$$

Example: Let us evaluate the integral along the line segment from $(0,0)$ to $(1,1)$.

Parameterization:
$$\vec{r}(t) = \langle t, t \rangle \quad \text{for} \quad 0 \leq t \leq 1.$$

Compute derivatives:
$$\frac{dx}{dt} = 1, \quad \frac{dy}{dt} = 1.$$

Substitute into the integral:

$$\int_0^1 \left(t^2 \cdot 1 + t^2 \cdot 1 \right) dt = \int_0^1 2t^2 \, dt = 2 \left[\frac{t^3}{3} \right]_0^1 = 2 \cdot \frac{1}{3} = \frac{2}{3}.$$

Thus, the value of the line integral along the straight path from $(0,0)$ to $(1,1)$ is $\frac{2}{3}$.

---

if there exists a scalar potential function $\phi(x, y, z)$ such that:

$$F = \nabla\phi$$

In other words:

$$P = \frac{\partial\phi}{\partial x}, \quad Q = \frac{\partial\phi}{\partial y}, \quad R = \frac{\partial\phi}{\partial z}$$

A key property of conservative vector fields is that their curl must be zero:

$$\nabla \times F = 0$$

Application to the Given Vector Field

Given:

$$F = y\hat{i} - x\hat{j}$$

This vector field has components:

$$P(x, y) = y, \quad Q(x, y) = -x, \quad R(x, y, z) = 0$$

Since the vector field is two-dimensional (no $\hat{k}$ component), we can simplify our analysis by considering only the $\hat{i}$ and $\hat{j}$ components.

Checking the Curl of F

For a two-dimensional vector field $F = P(x, y)\hat{i} + Q(x, y)\hat{j}$, the curl is given by:

$$\nabla \times F = \left(\frac{\partial Q}{\partial x} - \frac{\partial P}{\partial y}\right)\hat{k}$$

Compute the partial derivatives:

$$\frac{\partial Q}{\partial x} = \frac{\partial(-x)}{\partial x} = -1$$

$$\frac{\partial P}{\partial y} = \frac{\partial y}{\partial y} = 1$$

Thus:

$$\nabla \times F = (-1 - 1)\hat{k} = -2\hat{k} \neq 0$$

Since the curl of F is not zero, the vector field F is **not** conservative.

Alternative Verification Using Path Independence

Another characterization of conservative vector fields is that the line integral between two points is path-independent. To further substantiate our conclusion, consider evaluating the line integral of F along two different paths between the same points.

Path 1: Along the x-axis from $(0,0)$ to $(a,0)$, then parallel to the y-axis to (a,b)

Parametrize the path in two segments:

$$\text{Segment 1: } \gamma_1(t) = (t,0), \quad t \in [0,a]$$

$$\text{Segment 2: } \gamma_2(t) = (a,t), \quad t \in [0,b]$$

Compute the line integral:

$$\int_\gamma \boldsymbol{F} \cdot d\boldsymbol{r} = \int_{\gamma_1} y\,dx - x\,dy + \int_{\gamma_2} y\,dx - x\,dy$$

For γ_1:

$$y = 0 \implies \int_{\gamma_1} 0 \cdot dx - t \cdot 0 = 0$$

For γ_2:

$$dx = 0, \quad dy = dt \implies \int_{\gamma_2} t \cdot 0 - a \cdot dt = -a \int_0^b dt = -ab$$

Thus, the total line integral along Path 1 is:

$$\int_\gamma \boldsymbol{F} \cdot d\boldsymbol{r} = 0 - ab = -ab$$

Path 2: Along the y-axis from $(0,0)$ to $(0,b)$, then parallel to the x-axis to (a,b)

Parametrize the path in two segments:

$$\text{Segment 1: } \gamma_3(t) = (0,t), \quad t \in [0,b]$$

$$\text{Segment 2: } \gamma_4(t) = (t,b), \quad t \in [0,a]$$

Compute the line integral:

$$\int_\gamma \boldsymbol{F} \cdot d\boldsymbol{r} = \int_{\gamma_3} y\,dx - x\,dy + \int_{\gamma_4} y\,dx - x\,dy$$

For γ_3:

$$x = 0 \implies \int_{\gamma_3} t \cdot 0 - 0 \cdot dt = 0$$

For γ_4:

$$y = b, \quad dy = 0 \implies \int_{\gamma_4} b \cdot dt - t \cdot 0 = b \int_0^a dt = ba$$

Thus, the total line integral along Path 2 is:

$$\int_\gamma F \cdot dr = 0 + ab = ab$$

Conclusion from Path Integrals

The line integral of F along Path 1 is $-ab$ and along Path 2 is ab. Since the integrals are different for the same endpoints, the line integral is path-dependent. This confirms that F is not a conservative vector field.

Final Conclusion

Based on the evaluation of the curl of F and the path dependence of the line integral, we conclude that the vector field $F = y\hat{i} - x\hat{j}$ is **not** conservative.

$$\boxed{\text{The vector field } F = y\hat{i} - x\hat{j} \text{ is not conservative.}}$$

9.8 Compute the curl of $F = (yz)\hat{i} + (xz)\hat{j} + (xy)\hat{k}$.

To compute the curl of the vector field $F = (yz)\hat{i} + (xz)\hat{j} + (xy)\hat{k}$, we utilize the definition of the curl in three-dimensional Cartesian coordinates. The curl of a vector field $F = P\hat{i} + Q\hat{j} + R\hat{k}$ is given by:

$$\nabla \times F = \left(\frac{\partial R}{\partial y} - \frac{\partial Q}{\partial z}\right)\hat{i} + \left(\frac{\partial P}{\partial z} - \frac{\partial R}{\partial x}\right)\hat{j} + \left(\frac{\partial Q}{\partial x} - \frac{\partial P}{\partial y}\right)\hat{k}$$

Given:

$$P = yz, \quad Q = xz, \quad R = xy$$

Let's compute each component of the curl separately.

- Compute the $\hat{i}$-component: $\frac{\partial R}{\partial y} - \frac{\partial Q}{\partial z}$

$$\frac{\partial R}{\partial y} = \frac{\partial(xy)}{\partial y} = x$$

$$\frac{\partial Q}{\partial z} = \frac{\partial(xz)}{\partial z} = x$$

$$\frac{\partial R}{\partial y} - \frac{\partial Q}{\partial z} = x - x = 0$$

- Compute the $\hat{j}$-component: $\frac{\partial P}{\partial z} - \frac{\partial R}{\partial x}$

$$\frac{\partial P}{\partial z} = \frac{\partial(yz)}{\partial z} = y$$

$$\frac{\partial R}{\partial x} = \frac{\partial(xy)}{\partial x} = y$$

$$\frac{\partial P}{\partial z} - \frac{\partial R}{\partial x} = y - y = 0$$

- Compute the $\hat{k}$-component: $\frac{\partial Q}{\partial x} - \frac{\partial P}{\partial y}$

$$\frac{\partial Q}{\partial x} = \frac{\partial(xz)}{\partial x} = z$$

$$\frac{\partial P}{\partial y} = \frac{\partial(yz)}{\partial y} = z$$

$$\frac{\partial Q}{\partial x} - \frac{\partial P}{\partial y} = z - z = 0$$

Combining the computed components, we find:

$$\nabla \times \boldsymbol{F} = 0\,\hat{i} + 0\,\hat{j} + 0\,\hat{k} = \boldsymbol{0}$$

Therefore, the curl of F is the zero vector.

$$\nabla \times \boldsymbol{F} = \boldsymbol{0}$$

9.9 Find the divergence of $F = e^x\hat{i} + e^y\hat{j} + e^z\hat{k}$.

To find the divergence of the vector field $\boldsymbol{F} = e^x\hat{i} + e^y\hat{j} + e^z\hat{k}$, we utilize the definition of divergence in three-dimensional Cartesian coordinates. The divergence of a vector field $\boldsymbol{F} = P\hat{i} + Q\hat{j} + R\hat{k}$ is given by:

$$\nabla \cdot \boldsymbol{F} = \frac{\partial P}{\partial x} + \frac{\partial Q}{\partial y} + \frac{\partial R}{\partial z}$$

Where:

$$P = e^x$$
$$Q = e^y$$
$$R = e^z$$

Let's compute each partial derivative step by step.

Step 1: Compute $\frac{\partial P}{\partial x}$

$$\frac{\partial P}{\partial x} = \frac{\partial}{\partial x}\left(e^x\right) = e^x$$

Step 2: Compute $\frac{\partial Q}{\partial y}$

$$\frac{\partial Q}{\partial y} = \frac{\partial}{\partial y}\left(e^y\right) = e^y$$

Step 3: Compute $\frac{\partial R}{\partial z}$

$$\frac{\partial R}{\partial z} = \frac{\partial}{\partial z}\left(e^z\right) = e^z$$

Step 4: Sum the Partial Derivatives Add the results from the three partial derivatives to find the divergence:

$$\nabla \cdot F = e^x + e^y + e^z$$

The divergence of the vector field $F = e^x \hat{i} + e^y \hat{j} + e^z \hat{k}$ is:

$$\nabla \cdot F = e^x + e^y + e^z$$

This scalar function represents the rate at which the vector field F is "spreading out" from a given point in space.

Interpretation Since all components of F are exponential functions of their respective variables, the divergence indicates that the vector field is expanding uniformly in all directions. The positive divergence at any point implies a source-like behavior, where vectors are emanating outward from that point.

Applications Understanding the divergence of a vector field is crucial in various applications, such as fluid dynamics, electromagnetism, and differential geometry. For example, in fluid dynamics, a positive divergence at a point indicates a source of fluid, whereas a negative divergence would indicate a sink.

Verification Using Components To ensure the correctness of our computation, let's verify each component:

- $\frac{\partial P}{\partial x} = \frac{\partial e^x}{\partial x} = e^x$

- $\frac{\partial Q}{\partial y} = \frac{\partial e^y}{\partial y} = e^y$

- $\frac{\partial R}{\partial z} = \frac{\partial e^z}{\partial z} = e^z$

189

Adding these results confirms our earlier conclusion:

$$\nabla \cdot \boldsymbol{F} = e^x + e^y + e^z$$

Final Answer

$$\nabla \cdot \boldsymbol{F} = e^x + e^y + e^z$$

9.10 Solve for the potential function ϕ such that $\boldsymbol{F} = \nabla \phi$

To find the potential function ϕ for a conservative vector field $\boldsymbol{F}$ such that $\boldsymbol{F} = \nabla \phi$, we follow a systematic approach. This involves integrating the components of $\boldsymbol{F}$ and ensuring consistency among them.

- **Given Vector Field**

Assume the vector field $\boldsymbol{F}$ is defined as:

$$\boldsymbol{F} = \langle M(x,y,z),\ N(x,y,z),\ P(x,y,z) \rangle$$

where M, N, and P are the components of $\boldsymbol{F}$ in the x, y, and z directions, respectively.

- **Condition for Conservativeness**

For $\boldsymbol{F}$ to be conservative, it must satisfy the condition:

$$\frac{\partial M}{\partial y} = \frac{\partial N}{\partial x}, \quad \frac{\partial M}{\partial z} = \frac{\partial P}{\partial x}, \quad \frac{\partial N}{\partial z} = \frac{\partial P}{\partial y}$$

These are the **cross partial derivative conditions** derived from the equality of mixed partials of ϕ.

- **Finding the Potential Function ϕ**

Assuming $\boldsymbol{F}$ is conservative and the conditions above are satisfied, we can find ϕ by integrating the components of $\boldsymbol{F}$.

- **Step 1: Integrate M with respect to x**

190

First, integrate the x-component of $\boldsymbol{F}$ with respect to x:

$$\phi(x, y, z) = \int M(x, y, z)\, dx + h(y, z)$$

where $h(y, z)$ is an arbitrary function of y and z arising from the integration.

- **Step 2: Differentiate ϕ with respect to y and Equate to N**

Differentiate the obtained ϕ with respect to y:

$$\frac{\partial\phi}{\partial y} = \int \frac{\partial M}{\partial y}\, dx + \frac{\partial h}{\partial y} = N(x, y, z)$$

From this equation, we can solve for $\frac{\partial h}{\partial y}$:

$$\frac{\partial h}{\partial y} = N(x, y, z) - \frac{\partial}{\partial y}\left(\int M(x, y, z)\, dx\right)$$

Integrate $\frac{\partial h}{\partial y}$ with respect to y to find $h(y, z)$:

$$h(y, z) = \int \left[N(x, y, z) - \frac{\partial}{\partial y}\left(\int M(x, y, z)\, dx\right)\right] dy + g(z)$$

where $g(z)$ is an arbitrary function of z.

- **Step 3: Differentiate ϕ with respect to z and Equate to P**

Finally, differentiate ϕ with respect to z:

$$\frac{\partial\phi}{\partial z} = \frac{\partial}{\partial z}\left(\int M\, dx\right) + \frac{\partial h}{\partial z} = P(x, y, z)$$

From this equation, solve for $\frac{\partial h}{\partial z}$:

$$\frac{\partial h}{\partial z} = P(x, y, z) - \frac{\partial}{\partial z}\left(\int M(x, y, z)\, dx\right)$$

Integrate $\frac{\partial h}{\partial z}$ with respect to z to determine $g(z)$:

$$g(z) = \int \left[P(x, y, z) - \frac{\partial}{\partial z}\left(\int M(x, y, z)\, dx\right)\right] dz + C$$

where C is a constant of integration.

- **Complete Expression for** ϕ

Substituting $h(y, z)$ and $g(z)$ back into the expression for ϕ, we obtain:

$$\phi(x, y, z) = \int M(x, y, z)\, dx + \int \left[N(x, y, z) - \frac{\partial}{\partial y} \left(\int M\, dx \right) \right] dy$$

$$+ \int \left[P(x, y, z) - \frac{\partial}{\partial z} \left(\int M\, dx \right) \right] dz + C$$

- **Example**

To illustrate the process, consider a specific vector field:

$$\boldsymbol{F} = \langle 2xy, x^2 + 3z^2, 6zy \rangle$$

- **Check Conservativeness**

First, verify that $\boldsymbol{F}$ is conservative by checking the cross partial derivatives:

$$\frac{\partial M}{\partial y} = 2x, \quad \frac{\partial N}{\partial x} = 2x \quad \Rightarrow \quad \frac{\partial M}{\partial y} = \frac{\partial N}{\partial x}$$

$$\frac{\partial M}{\partial z} = 0, \quad \frac{\partial P}{\partial x} = 0 \quad \Rightarrow \quad \frac{\partial M}{\partial z} = \frac{\partial P}{\partial x}$$

$$\frac{\partial N}{\partial z} = 6z, \quad \frac{\partial P}{\partial y} = 6z \quad \Rightarrow \quad \frac{\partial N}{\partial z} = \frac{\partial P}{\partial y}$$

All conditions are satisfied, so $\boldsymbol{F}$ is conservative.

- **Integrate M with respect to x**

$$\phi(x, y, z) = \int 2xy\, dx = x^2 y + h(y, z)$$

- **Differentiate ϕ with respect to y and Equate to N**

$$\frac{\partial \phi}{\partial y} = x^2 + \frac{\partial h}{\partial y} = x^2 + 3z^2$$

$$\Rightarrow \frac{\partial h}{\partial y} = 3z^2$$

Integrate with respect to y:

$$h(y, z) = 3z^2 y + g(z)$$

- **Differentiate ϕ with respect to z and Equate to P**

$$\frac{\partial\phi}{\partial z} = 6zy + \frac{dg}{dz} = 6zy$$

$$\Rightarrow \frac{dg}{dz} = 0 \quad \Rightarrow \quad g(z) = C$$

- **Final Potential Function ϕ**

Substituting $h(y, z)$ and $g(z)$ back:

$$\phi(x, y, z) = x^2 y + 3z^2 y + C$$

where C is the constant of integration.

- **Verification**

To ensure the correctness of ϕ, compute $\nabla\phi$:

$$\nabla\phi = \left\langle \frac{\partial\phi}{\partial x}, \frac{\partial\phi}{\partial y}, \frac{\partial\phi}{\partial z} \right\rangle = \langle 2xy, \; x^2 + 3z^2, \; 6zy \rangle = \boldsymbol{F}$$

This confirms that ϕ is indeed the potential function for $\boldsymbol{F}$.

By systematically integrating the components of a conservative vector field and ensuring consistency among the partial derivatives, we have successfully determined the potential function ϕ such that $\boldsymbol{F} = \nabla\phi$. This process is fundamental in vector calculus, particularly in fields like physics and engineering, where potential functions play a crucial role in describing conservative forces and energy fields.

Chapter 10

Differential Equations

This chapter examines differential equations, presenting problems that explore various methods for solving ordinary differential equations (ODEs). Topics include first-order linear and nonlinear ODEs, higher-order linear ODEs, and systems of differential equations. The problems are designed to enhance understanding of the techniques used to model and solve dynamic systems. Solutions are provided with detailed explanations, aiding in the acquisition of the skills necessary for analyzing and solving differential equations in both theoretical and applied contexts.

10.1 Solve $\frac{dy}{dx} = ky$ with initial condition $y(0) = y_0$.

We are tasked with solving the first-order linear ordinary differential equation (ODE):

$$\frac{dy}{dx} = ky$$

where k is a constant, subject to the initial condition $y(0) = y_0$.

- **Step 1: Separation of Variables**

To solve the ODE, we begin by separating the variables y and x. This involves rearranging the equation so that all terms involving y are on one side and all terms involving x are on the other side.

$$\frac{dy}{dx} = ky \quad \Rightarrow \quad \frac{dy}{y} = k\,dx$$

- **Step 2: Integration**

Next, we integrate both sides of the equation to find the general solution.

$$\int \frac{1}{y}\,dy = \int k\,dx$$

Performing the integrations:

$$\ln|y| = kx + C$$

where C is the constant of integration.

- **Step 3: Solving for y**

To express y explicitly, we exponentiate both sides of the equation to eliminate the natural logarithm.

$$e^{\ln|y|} = e^{kx+C} \quad \Rightarrow \quad |y| = e^{kx} \cdot e^{C}$$

Since e^{C} is a positive constant, we can denote it as A (where $A > 0$). Additionally, we can drop the absolute value since A can accommodate the sign of y.

$$y = Ae^{kx}$$

- **Step 4: Applying the Initial Condition**

We now apply the initial condition $y(0) = y_0$ to determine the constant A.

$$y(0) = Ae^{k \cdot 0} = Ae^{0} = A \cdot 1 = A$$

Thus,

$$A = y_0$$

- **Final Solution**

Substituting $A = y_0$ back into the general solution, we obtain the particular solution satisfying the initial condition:

$$y(x) = y_0 e^{kx}$$

The solution to the differential equation $\frac{dy}{dx} = ky$ with the initial condition $y(0) = y_0$ is:

$$y(x) = y_0 e^{kx}$$

This exponential solution describes how y evolves with respect to x under the influence of the constant rate k.

10.2 Find the general solution to $\frac{dy}{dx} + y \tan(x) = \sin(x)$.

We are tasked with finding the general solution to the first-order linear ordinary differential equation (ODE):

$$\frac{dy}{dx} + y \tan(x) = \sin(x)$$

Standard Form

First, we rewrite the ODE in the standard linear form:

$$\frac{dy}{dx} + P(x)y = Q(x)$$

where:
$$P(x) = \tan(x) \quad \text{and} \quad Q(x) = \sin(x)$$

Integrating Factor

To solve this linear ODE, we employ the method of integrating factors. The integrating factor, $\mu(x)$, is given by:

$$\mu(x) = e^{\int P(x)\, dx} = e^{\int \tan(x)\, dx}$$

We compute the integral:

$$\int \tan(x)\, dx = -\ln|\cos(x)| + C$$

Ignoring the constant of integration C (since it will cancel out in the integrating factor), we have:

$$\mu(x) = e^{-\ln|\cos(x)|} = \frac{1}{\cos(x)} = \sec(x)$$

Multiplying by the Integrating Factor

We multiply both sides of the ODE by the integrating factor $\mu(x) = \sec(x)$:

$$\sec(x)\frac{dy}{dx} + y\sec(x)\tan(x) = \sin(x)\sec(x)$$

Simplifying the right-hand side:

$$\sin(x)\sec(x) = \frac{\sin(x)}{\cos(x)} = \tan(x)$$

Thus, the equation becomes:

$$\sec(x)\frac{dy}{dx} + y\sec(x)\tan(x) = \tan(x)$$

Recognizing the Left-Hand Side as a Derivative

Notice that the left-hand side of the equation is the derivative of $y\sec(x)$ with respect to x:

$$\frac{d}{dx}\left(y\sec(x)\right) = \sec(x)\frac{dy}{dx} + y\sec(x)\tan(x)$$

Therefore, we can rewrite the ODE as:

198

$$\frac{d}{dx}\left(y\sec(x)\right) = \tan(x)$$

Integrating Both Sides

Integrate both sides with respect to x:

$$\int \frac{d}{dx}\left(y\sec(x)\right)\,dx = \int \tan(x)\,dx$$

The left-hand side simplifies to:

$$y\sec(x) = -\ln|\cos(x)| + C$$

where C is the constant of integration.

Solving for y

Finally, solve for y by multiplying both sides by $\cos(x)$:

$$y = \cos(x)\left(-\ln|\cos(x)| + C\right)$$

Simplifying, we obtain the general solution:

$$y(x) = -\cos(x)\ln|\cos(x)| + C\cos(x)$$

where C is an arbitrary constant.

Final Answer

The general solution to the differential equation $\frac{dy}{dx} + y\tan(x) = \sin(x)$ is:

$$y(x) = -\cos(x)\ln|\cos(x)| + C\cos(x)$$

where C is an arbitrary constant.

199

10.3 Solve the homogeneous differential equation $y'' + y = 0$.

We are tasked with solving the second-order homogeneous linear differential equation:

$$y'' + y = 0$$

This differential equation features constant coefficients and is linear, making it amenable to standard solution techniques. The general approach involves finding the characteristic equation associated with the differential equation, determining its roots, and then constructing the general solution based on these roots.

Formulating the Characteristic Equation

A linear differential equation with constant coefficients of the form:

$$ay'' + by' + cy = 0$$

can be solved by assuming a solution of the form $y = e^{rt}$, where r is a constant to be determined. Substituting $y = e^{rt}$ into the differential equation, we obtain:

$$ar^2 e^{rt} + bre^{rt} + ce^{rt} = 0$$

Dividing both sides by e^{rt} (which is never zero), the characteristic equation is:

$$ar^2 + br + c = 0$$

Applying this to our specific equation $y'' + y = 0$:

$$1 \cdot y'' + 0 \cdot y' + 1 \cdot y = 0$$

Here, the coefficients are:

$$a = 1, \quad b = 0, \quad c = 1$$

Thus, the characteristic equation becomes:

$$r^2 + 1 = 0$$

Solving the Characteristic Equation

The characteristic equation to solve is:

$$r^2 + 1 = 0$$

Solving for r:

$$r^2 = -1 \; r = \pm\sqrt{-1} \; r = \pm i$$

The roots of the characteristic equation are complex conjugates: $r = i$ and $r = -i$.

Constructing the General Solution

When the characteristic equation yields complex roots of the form $r = \alpha \pm \beta i$, the general solution to the differential equation is given by:

$$y(t) = e^{\alpha t} \left(C_1 \cos(\beta t) + C_2 \sin(\beta t) \right)$$

In our case, the roots are $r = \pm i$, which implies:

$$\alpha = 0, \quad \beta = 1$$

Substituting these values into the general solution formula:

$$y(t) = e^{0 \cdot t} \left(C_1 \cos(1 \cdot t) + C_2 \sin(1 \cdot t) \right) \; y(t) = C_1 \cos(t) + C_2 \sin(t)$$

Here, C_1 and C_2 are arbitrary constants determined by initial conditions or boundary values.

The general solution to the homogeneous differential equation $y'' + y = 0$ is:

$$y(t) = C_1 \cos(t) + C_2 \sin(t)$$

This solution represents a combination of the fundamental solutions $\cos(t)$ and $\sin(t)$, which are linearly independent. The constants C_1 and C_2 can be determined if additional information, such as initial conditions, is provided.

Verification of the Solution

To ensure the correctness of the solution, we can verify by substituting $y(t) = C_1 \cos(t) + C_2 \sin(t)$ back into the original differential equation.

First, compute the first and second derivatives of $y(t)$:

$$y'(t) = -C_1 \sin(t) + C_2 \cos(t) \quad y''(t) = -C_1 \cos(t) - C_2 \sin(t)$$

Substitute $y(t)$ and $y''(t)$ into the differential equation:

$$y'' + y = (-C_1 \cos(t) - C_2 \sin(t)) + (C_1 \cos(t) + C_2 \sin(t)) = 0$$

This confirms that the solution satisfies the differential equation.

Final Remarks

The method of characteristic equations is a powerful tool for solving linear differential equations with constant coefficients. In cases where the characteristic equation yields complex roots, as demonstrated, the solutions involve sine and cosine functions, reflecting oscillatory behavior characteristic of such differential systems.

10.4 Find the particular solution to $y'' - 3y' + 2y = 0$ given initial conditions

To find the particular solution to the differential equation

$$y'' - 3y' + 2y = 0,$$

we will follow a systematic approach involving the characteristic equation method. This method is suitable for solving linear homogeneous ordinary differential equations with constant coefficients.

- **Step 1: Write the Characteristic Equation**

The given differential equation is

$$y'' - 3y' + 2y = 0.$$

Assuming a solution of the form $y = e^{rx}$, where r is a constant to be determined, we substitute into the differential equation.

First, compute the derivatives:

$$y' = re^{rx}, \quad y'' = r^2 e^{rx}.$$

Substituting these into the differential equation gives:

$$r^2 e^{rx} - 3re^{rx} + 2e^{rx} = 0.$$

Factoring out e^{rx} (which is never zero) leads to the characteristic equation:

$$r^2 - 3r + 2 = 0.$$

- **Step 2: Solve the Characteristic Equation**

The characteristic equation is a quadratic equation:

$$r^2 - 3r + 2 = 0.$$

We can solve for r using the quadratic formula:

$$r = \frac{3 \pm \sqrt{(-3)^2 - 4 \cdot 1 \cdot 2}}{2 \cdot 1} = \frac{3 \pm \sqrt{9 - 8}}{2} = \frac{3 \pm 1}{2}.$$

This yields two distinct real roots:

$$r_1 = \frac{3 + 1}{2} = 2, \quad r_2 = \frac{3 - 1}{2} = 1.$$

- **Step 3: Write the General Solution**

Since we have two distinct real roots, the general solution to the differential equation is:

$$y(x) = C_1 e^{r_1 x} + C_2 e^{r_2 x} = C_1 e^{2x} + C_2 e^x,$$

where C_1 and C_2 are arbitrary constants determined by initial conditions.

- **Step 4: Apply Initial Conditions to Find Particular Solution**

To find the particular solution, specific initial conditions are required. Suppose we are given:

$$y(x_0) = y_0, \quad y'(x_0) = y_0'.$$

Using these, we can set up a system of equations to solve for C_1 and C_2.

- **Example**

203

Assume the initial conditions are:

$$y(0) = y_0, \quad y'(0) = y_0'.$$

Substitute $x = 0$ into the general solution:

$$y(0) = C_1 e^0 + C_2 e^0 = C_1 + C_2 = y_0.$$

Differentiate the general solution:

$$y'(x) = 2C_1 e^{2x} + C_2 e^x.$$

Substitute $x = 0$:

$$y'(0) = 2C_1 + C_2 = y_0'.$$

Now, we have the system of equations:

$$\begin{cases} C_1 + C_2 = y_0, \\ 2C_1 + C_2 = y_0'. \end{cases}$$

Subtract the first equation from the second:

$$(2C_1 + C_2) - (C_1 + C_2) = y_0' - y_0 \Rightarrow C_1 = y_0' - y_0.$$

Substitute C_1 back into the first equation:

$$(y_0' - y_0) + C_2 = y_0 \Rightarrow C_2 = 2y_0 - y_0'.$$

Thus, the particular solution is:

$$y(x) = (y_0' - y_0)e^{2x} + (2y_0 - y_0')e^x.$$

The particular solution to the differential equation $y'' - 3y' + 2y = 0$ is determined by the initial conditions provided. By solving the characteristic equation, we obtain the general solution. Applying the initial conditions allows us to solve for the constants C_1 and C_2, thereby yielding the particular solution specific to the given scenario.

- **Final Particular Solution**

If specific initial conditions are provided, substitute them into the general solution as demonstrated in the example above to find the particular solution unique to those conditions.

$$y(x) = C_1 e^{2x} + C_2 e^x,$$

where C_1 and C_2 are constants determined by the initial conditions.

10.5 Determine the solution to $\frac{dy}{dx} = \frac{x+y}{x-y}$.

To solve the differential equation:

$$\frac{dy}{dx} = \frac{x+y}{x-y}$$

we can employ a substitution method to simplify the equation. The equation is a first-order ordinary differential equation (ODE) and appears to be homogeneous. Let's proceed with the following steps:

Step 1: Identify Homogeneity

A differential equation is homogeneous if it can be expressed as a function of $\frac{y}{x}$. Observe that both the numerator and the denominator on the right-hand side are linear combinations of x and y, which suggests that the equation is homogeneous.

Step 2: Make a Substitution

Let:

$$v = \frac{y}{x}$$

Then:

$$y = vx$$

Differentiate both sides with respect to x:

$$\frac{dy}{dx} = v + x\frac{dv}{dx}$$

Step 3: Substitute into the Original Equation

Substituting $y = vx$ and $\frac{dy}{dx} = v + x\frac{dv}{dx}$ into the original equation:

$$v + x\frac{dv}{dx} = \frac{x + vx}{x - vx} = \frac{1+v}{1-v}$$

205

Simplify:

$$v + x\frac{dv}{dx} = \frac{1+v}{1-v}$$

Step 4: Rearrange the Equation

Isolate the term involving $\frac{dv}{dx}$:

$$x\frac{dv}{dx} = \frac{1+v}{1-v} - v = \frac{1+v-v(1-v)}{1-v} = \frac{1+v-v+v^2}{1-v} = \frac{1+v^2}{1-v}$$

Thus, we have:

$$\frac{dv}{dx} = \frac{1+v^2}{x(1-v)}$$

Step 5: Separate Variables

Rearrange the equation to separate the variables v and x:

$$\frac{1-v}{1+v^2}\,dv = \frac{1}{x}\,dx$$

Step 6: Integrate Both Sides

Integrate both sides:

$$\int \frac{1-v}{1+v^2}\,dv = \int \frac{1}{x}\,dx$$

Break down the left integral:

$$\int \frac{1-v}{1+v^2}\,dv = \int \frac{1}{1+v^2}\,dv - \int \frac{v}{1+v^2}\,dv$$

Calculate the integrals:

$$\int \frac{1}{1+v^2} \, dv = \arctan v$$

$$\int \frac{v}{1+v^2} \, dv = \frac{1}{2} \ln|1+v^2|$$

Thus, the left side becomes:

$$\arctan v - \frac{1}{2} \ln|1+v^2| = \ln|x| + C$$

Where C is the constant of integration.

Step 7: Substitute Back $v = \frac{y}{x}$

Replace v with $\frac{y}{x}$:

$$\arctan\left(\frac{y}{x}\right) - \frac{1}{2} \ln\left(1 + \left(\frac{y}{x}\right)^2\right) = \ln|x| + C$$

Step 8: Simplify the Equation

Simplify the logarithmic term:

$$\ln\left(1 + \frac{y^2}{x^2}\right) = \ln\left(\frac{x^2+y^2}{x^2}\right) = \ln(x^2+y^2) - \ln(x^2)$$

Substituting back:

$$\arctan\left(\frac{y}{x}\right) - \frac{1}{2}\left[\ln(x^2+y^2) - \ln(x^2)\right] = \ln|x| + C$$

Simplify further:

$$\arctan\left(\frac{y}{x}\right) - \frac{1}{2}\ln(x^2+y^2) + \frac{1}{2}\ln(x^2) = \ln|x| + C$$

$$\arctan\left(\frac{y}{x}\right) - \frac{1}{2}\ln(x^2+y^2) = \ln|x| + C - \frac{1}{2}\ln(x^2)$$

Since $\ln(x^2) = 2\ln|x|$, we have:

207

$$\arctan\left(\frac{y}{x}\right) - \frac{1}{2}\ln(x^2 + y^2) = -\ln|x| + C$$

We can absorb constants into C and rearrange:

$$\arctan\left(\frac{y}{x}\right) - \frac{1}{2}\ln(x^2 + y^2) + \ln|x| = C$$

Step 9: General Solution

The general solution to the differential equation is:

$$\arctan\left(\frac{y}{x}\right) - \frac{1}{2}\ln(x^2 + y^2) + \ln|x| = C$$

Where C is an arbitrary constant. This implicit solution relates y and x and encompasses all possible solutions to the original differential equation.

By recognizing the homogeneity of the differential equation and applying an appropriate substitution, we successfully derived the general solution. The integration involved standard techniques, and the final implicit solution captures the relationship between y and x for the given ODE.

10.6 Solve the system $\frac{dx}{dt} = 3x + 4y$, $\frac{dy}{dt} = -4x + 3y$.

To solve the system of differential equations:

$$\frac{dx}{dt} = 3x + 4y, \quad \frac{dy}{dt} = -4x + 3y,$$

we can employ the method of eigenvalues and eigenvectors, which is particularly effective for linear systems with constant coefficients. Below are the detailed steps to find the general solution.

- **Step 1: Write the System in Matrix Form** First, express the system in matrix form:

$$\boldsymbol{X}' = A\boldsymbol{X},$$

where

208

$$X = \begin{pmatrix} x \\ y \end{pmatrix}, \quad A = \begin{pmatrix} 3 & 4 \\ -4 & 3 \end{pmatrix}.$$

- **Step 2: Find the Eigenvalues of Matrix** A The eigenvalues λ are found by solving the characteristic equation:

$$\det(A - \lambda I) = 0.$$

Compute $A - \lambda I$:

$$A - \lambda I = \begin{pmatrix} 3 - \lambda & 4 \\ -4 & 3 - \lambda \end{pmatrix}.$$

Compute the determinant:

$$\det(A - \lambda I) = (3 - \lambda)(3 - \lambda) - (4)(-4) = (3 - \lambda)^2 + 16.$$

Set the determinant equal to zero:

$$(3 - \lambda)^2 + 16 = 0 \Rightarrow (3 - \lambda)^2 = -16.$$

Solve for λ:

$$3 - \lambda = \pm 4i \Rightarrow \lambda = 3 \pm 4i.$$

Thus, the eigenvalues are:

$$\lambda_{1,2} = 3 \pm 4i.$$

- **Step 3: Find the Eigenvectors Corresponding to Each Eigenvalue** Let's find the eigenvector corresponding to $\lambda = 3 + 4i$. The process for $\lambda = 3 - 4i$ will be analogous.
Solve $(A - \lambda I)v = 0$:

$$\begin{pmatrix} 3 - (3 + 4i) & 4 \\ -4 & 3 - (3 + 4i) \end{pmatrix} \begin{pmatrix} v_1 \\ v_2 \end{pmatrix} = \begin{pmatrix} 0 \\ 0 \end{pmatrix}.$$

Simplify the matrix:

209

$$\begin{pmatrix} -4i & 4 \\ -4 & -4i \end{pmatrix} \begin{pmatrix} v_1 \\ v_2 \end{pmatrix} = \begin{pmatrix} 0 \\ 0 \end{pmatrix}.$$

This system reduces to:

$$-4iv_1 + 4v_2 = 0 \Rightarrow v_2 = \frac{i}{1} v_1.$$

Let $v_1 = 1$, then $v_2 = i$. Thus, an eigenvector corresponding to $\lambda = 3 + 4i$ is:

$$v = \begin{pmatrix} 1 \\ i \end{pmatrix}.$$

- **Step 4: Construct the General Solution** For complex eigenvalues $\lambda = \alpha \pm \beta i$ with eigenvectors $v = a \pm ib$, the general solution is:

$$X(t) = e^{\alpha t} \left[a \cos(\beta t) - b \sin(\beta t) \right] + e^{\alpha t} \left[b \cos(\beta t) + a \sin(\beta t) \right].$$

In our case, $\alpha = 3$, $\beta = 4$, and $v = \begin{pmatrix} 1 \\ i \end{pmatrix}$, so $a = \begin{pmatrix} 1 \\ 0 \end{pmatrix}$ and $b = \begin{pmatrix} 0 \\ 1 \end{pmatrix}$.

Therefore, the general solution is:

$$X(t) = e^{3t} \left[\begin{pmatrix} 1 \\ 0 \end{pmatrix} \cos(4t) - \begin{pmatrix} 0 \\ 1 \end{pmatrix} \sin(4t) \right] + e^{3t} \left[\begin{pmatrix} 0 \\ 1 \end{pmatrix} \cos(4t) + \begin{pmatrix} 1 \\ 0 \end{pmatrix} \sin(4t) \right]$$

$$= e^{3t} \begin{pmatrix} \cos(4t) + \sin(4t) \\ \cos(4t) - \sin(4t) \end{pmatrix}.$$

Thus, the solutions for $x(t)$ and $y(t)$ are:

$$\begin{cases} x(t) = e^{3t} \left(\cos(4t) + \sin(4t) \right), \\ y(t) = e^{3t} \left(\cos(4t) - \sin(4t) \right). \end{cases}$$

- **Summary of the Solution** The system of differential equations:

$$\frac{dx}{dt} = 3x + 4y, \quad \frac{dy}{dt} = -4x + 3y,$$

has the general solution:

$$\begin{cases} x(t) = e^{3t} \left(\cos(4t) + \sin(4t) \right), \\ y(t) = e^{3t} \left(\cos(4t) - \sin(4t) \right). \end{cases}$$

This solution is obtained by diagonalizing the system using eigenvalues and eigenvectors, and applying the theory of linear differential equations with constant coefficients.

- **Verification** To ensure the solution is correct, we can substitute $x(t)$ and $y(t)$ back into the original differential equations and verify that both equations are satisfied.

$$\begin{aligned} \frac{dx}{dt} &= \frac{d}{dt} \left[e^{3t} (\cos(4t) + \sin(4t)) \right] \\ &= e^{3t} \cdot 3(\cos(4t) + \sin(4t)) + e^{3t}(-4\sin(4t) + 4\cos(4t)) \\ &= e^{3t} [3\cos(4t) + 3\sin(4t) - 4\sin(4t) + 4\cos(4t)] \\ &= e^{3t} [7\cos(4t) - \sin(4t)] \\ &= 3x + 4y \quad \text{(as required)}. \end{aligned}$$

A similar verification can be performed for $\frac{dy}{dt}$.

Thus, the obtained solution satisfies the original system of differential equations.

10.7 Use Laplace transforms to solve $y'' + y = \delta(t - \pi)$.

To solve the differential equation

$$y''(t) + y(t) = \delta(t - \pi),$$

using Laplace transforms, we follow a systematic approach. Here, $\delta(t - \pi)$ represents the Dirac delta function centered at $t = \pi$. We'll assume that the initial conditions are zero, i.e.,

$$y(0) = 0 \quad \text{and} \quad y'(0) = 0.$$

This assumption is common unless specified otherwise.

- **Step 1: Take the Laplace Transform of Both Sides**

Recall that the Laplace transform of a function $f(t)$ is defined as

$$\mathcal{L}\{f(t)\} = \int_0^\infty e^{-st} f(t) dt.$$

Applying the Laplace transform to both sides of the differential equation:

$$\mathcal{L}\{y''(t) + y(t)\} = \mathcal{L}\{\delta(t - \pi)\}.$$

- **Step 2: Apply Linear Properties and Differentiation Rules**

Using the linearity of the Laplace transform:

$$\mathcal{L}\{y''(t)\} + \mathcal{L}\{y(t)\} = \mathcal{L}\{\delta(t - \pi)\}.$$

We know the following Laplace transforms:

$$\mathcal{L}\{y''(t)\} = s^2 Y(s) - sy(0) - y'(0),$$

$$\mathcal{L}\{y(t)\} = Y(s),$$

$$\mathcal{L}\{\delta(t - a)\} = e^{-as},$$

where $Y(s) = \mathcal{L}\{y(t)\}$.

Substituting the initial conditions $y(0) = 0$ and $y'(0) = 0$ into the equation:

$$s^2 Y(s) - 0 - 0 + Y(s) = e^{-\pi s}.$$

Simplifying:

$$(s^2 + 1)Y(s) = e^{-\pi s}.$$

- **Step 3: Solve for $Y(s)$**

$$Y(s) = \frac{e^{-\pi s}}{s^2 + 1}.$$

- **Step 4: Apply the Inverse Laplace Transform**

We need to find $y(t)$ such that:

$$y(t) = \mathcal{L}^{-1}\left\{\frac{e^{-\pi s}}{s^2 + 1}\right\}.$$

212

Recall the second shifting theorem (also known as the time-delay property) of Laplace transforms:

$$\mathcal{L}^{-1}\{e^{-as}F(s)\} = u(t - a)f(t - a),$$

where $u(t - a)$ is the Heaviside step function, and $f(t) = \mathcal{L}^{-1}\{F(s)\}$. Applying this theorem with $a = \pi$ and $F(s) = \frac{1}{s^2+1}$:

$$y(t) = u(t - \pi)\mathcal{L}^{-1}\left\{\frac{1}{s^2 + 1}\right\}(t - \pi).$$

We know that:

$$\mathcal{L}^{-1}\left\{\frac{1}{s^2 + 1}\right\} = \sin(t).$$

Therefore:

$$y(t) = u(t - \pi)\sin(t - \pi).$$

- **Step 5: Simplify the Solution**

The final solution incorporates the Heaviside function to account for the impulse at $t = \pi$:

$$y(t) = \begin{cases} 0, & t < \pi, \\ \sin(t - \pi), & t \geq \pi. \end{cases}$$

Alternatively, using the properties of the sine function:

$$y(t) = u(t - \pi)\sin(t - \pi).$$

The solution to the differential equation $y'' + y = \delta(t - \pi)$ with initial conditions $y(0) = 0$ and $y'(0) = 0$ is

$$y(t) = u(t - \pi)\sin(t - \pi),$$

which signifies that the system remains at rest for $t < \pi$ and then responds to the impulse at $t = \pi$ by oscillating sinusoidally.

- **Verification**

To ensure the correctness of the solution, let's verify that it satisfies the original differential equation.

For $t < \pi$:

$$y(t) = 0 \implies y''(t) + y(t) = 0 + 0 = 0.$$

213

The right-hand side is $\delta(t - \pi)$, which is zero for $t \neq \pi$. Hence, the equation holds.

For $t \geq \pi$:

$$y(t) = \sin(t - \pi),$$

$$y'(t) = \cos(t - \pi),$$

$$y''(t) = -\sin(t - \pi).$$

Substituting into the left-hand side:

$$y''(t) + y(t) = -\sin(t - \pi) + \sin(t - \pi) = 0.$$

At $t = \pi$, the delta function $\delta(t - \pi)$ contributes an impulse that causes the discontinuity in the derivative of $y(t)$, which is accounted for in the solution via the Heaviside function.

Thus, the solution satisfies the differential equation and the initial conditions.

10.8 Apply Euler's Method to approximate solutions to $\frac{dy}{dx} = x + y$.

Euler's Method is a numerical technique used to approximate solutions to ordinary differential equations (ODEs) when an analytical solution is difficult or impossible to obtain. It is particularly useful for solving initial value problems of the form:

$$\frac{dy}{dx} = f(x, y), \quad y(x_0) = y_0$$

Given the differential equation:

$$\frac{dy}{dx} = x + y$$

we aim to approximate the solution using Euler's Method. Below are the detailed steps involved in applying Euler's Method to this specific ODE.

Step 1: Identify the Components

First, identify the function $f(x, y)$ from the differential equation:

$$f(x, y) = x + y$$

Next, determine the initial condition $y(x_0) = y_0$. For the purpose of this example, let's assume:

$$x_0 = 0, \quad y_0 = 1$$

Step 2: Choose a Step Size h

The step size h determines the intervals at which we approximate the solution. A smaller h generally leads to more accurate results but requires more computations. For this example, let's choose:

$$h = 0.1$$

Step 3: Apply Euler's Formula

Euler's Method updates the solution using the following iterative formula:

$$y_{n+1} = y_n + h \cdot f(x_n, y_n)$$

Substituting $f(x, y) = x + y$:

$$y_{n+1} = y_n + h \cdot (x_n + y_n) = y_n + hx_n + hy_n$$

This can be further simplified to:

$$y_{n+1} = y_n(1 + h) + hx_n$$

Step 4: Iterative Computation

Using the initial condition $x_0 = 0$ and $y_0 = 1$, we compute subsequent values of y at each step.

215

Iteration 1: $n = 0$

$$x_0 = 0, \quad y_0 = 1$$

$$\begin{aligned} y_1 &= y_0 + h(x_0 + y_0) \\ &= 1 + 0.1(0 + 1) \\ &= 1 + 0.1 \\ &= 1.1 \end{aligned}$$

$$x_1 = x_0 + h = 0 + 0.1 = 0.1$$

Iteration 2: $n = 1$

$$x_1 = 0.1, \quad y_1 = 1.1$$

$$\begin{aligned} y_2 &= y_1 + h(x_1 + y_1) \\ &= 1.1 + 0.1(0.1 + 1.1) \\ &= 1.1 + 0.1 \times 1.2 \\ &= 1.1 + 0.12 = 1.22 \end{aligned}$$

$$x_2 = x_1 + h = 0.1 + 0.1 = 0.2$$

Iteration 3: $n = 2$

$$x_2 = 0.2, \quad y_2 = 1.22$$

$$\begin{aligned} y_3 &= y_2 + h(x_2 + y_2) \\ &= 1.22 + 0.1(0.2 + 1.22) \\ &= 1.22 + 0.1 \times 1.42 \\ &= 1.22 + 0.142 = 1.362 \end{aligned}$$

$$x_3 = x_2 + h = 0.2 + 0.1 = 0.3$$

Iteration 4: $n = 3$

$$x_3 = 0.3, \quad y_3 = 1.362$$

$$\begin{aligned}
y_4 &= y_3 + h(x_3 + y_3) \\
&= 1.362 + 0.1(0.3 + 1.362) \\
&= 1.362 + 0.1 \times 1.662 \\
&= 1.362 + 0.1662 = 1.5282
\end{aligned}$$

$$x_4 = x_3 + h = 0.3 + 0.1 = 0.4$$

Iteration 5: $n = 4$

$$x_4 = 0.4, \quad y_4 = 1.5282$$

$$\begin{aligned}
y_5 &= y_4 + h(x_4 + y_4) \\
&= 1.5282 + 0.1(0.4 + 1.5282) \\
&= 1.5282 + 0.1 \times 1.9282 \\
&= 1.5282 + 0.19282 = 1.72102
\end{aligned}$$

$$x_5 = x_4 + h = 0.4 + 0.1 = 0.5$$

Summary of Approximations

After five iterations, the approximate values of y at the corresponding x values are:

x_n	y_n
0.0	1.0000
0.1	1.1000
0.2	1.2200
0.3	1.3620
0.4	1.5282
0.5	1.7210

Euler's Method provides a straightforward procedure to approximate the solution of the differential equation $\frac{dy}{dx} = x + y$ using the initial condition $y(0) = 1$ and a step size $h = 0.1$. By iteratively applying the formula:

$$y_{n+1} = y_n + h(x_n + y_n)$$

we obtain successive approximations of y at discrete points x_n. While Euler's Method is simple to implement, its accuracy depends on the choice of step size h; smaller step sizes yield more accurate results but require more computations.

For improved accuracy, especially for larger intervals or more complex equations, higher-order numerical methods such as the Runge-Kutta methods may be employed.

10.9 Solve the boundary value problem $y'' + \lambda y = 0$ with $y(0) = 0, y(L) = 0$.

To solve the boundary value problem (BVP)

$$y'' + \lambda y = 0,$$

subject to the boundary conditions

$$y(0) = 0 \quad \text{and} \quad y(L) = 0,$$

we proceed by analyzing the differential equation and applying the boundary conditions to determine the permissible values of λ and the corresponding eigenfunctions $y(x)$.

General Solution of the Differential Equation

The differential equation is a second-order linear homogeneous ordinary differential equation (ODE) with constant coefficients. Its general solution depends on the sign of λ. We consider three cases: $\lambda > 0$, $\lambda = 0$, and $\lambda < 0$.

Case 1: $\lambda > 0$

Let $\lambda = \alpha^2$, where $\alpha > 0$. The ODE becomes:

$$y'' + \alpha^2 y = 0.$$

The characteristic equation associated with this ODE is:

$$r^2 + \alpha^2 = 0 \quad \Rightarrow \quad r = \pm i\alpha.$$

Thus, the general solution is:

$$y(x) = C_1 \cos(\alpha x) + C_2 \sin(\alpha x),$$

where C_1 and C_2 are arbitrary constants.

Case 2: $\lambda = 0$

When $\lambda = 0$, the ODE simplifies to:

$$y'' = 0.$$

Integrating twice, we obtain:

$$y(x) = C_1 x + C_2,$$

where C_1 and C_2 are constants of integration.

Case 3: $\lambda < 0$

Let $\lambda = -\beta^2$, where $\beta > 0$. The ODE becomes:

$$y'' - \beta^2 y = 0.$$

The characteristic equation is:

$$r^2 - \beta^2 = 0 \quad \Rightarrow \quad r = \pm \beta.$$

Thus, the general solution is:

$$y(x) = C_1 e^{\beta x} + C_2 e^{-\beta x},$$

where C_1 and C_2 are constants of integration.

Applying Boundary Conditions

We now apply the boundary conditions to each case to determine possible values of λ and corresponding eigenfunctions.

Case 1: $\lambda = \alpha^2 > 0$

The general solution is:

$$y(x) = C_1 \cos(\alpha x) + C_2 \sin(\alpha x).$$

Applying the boundary conditions:

219

Boundary Condition at $x = 0$:

$$y(0) = C_1 \cos(0) + C_2 \sin(0) = C_1 = 0.$$

Thus, $C_1 = 0$, and the solution reduces to:

$$y(x) = C_2 \sin(\alpha x).$$

Boundary Condition at $x = L$:

$$y(L) = C_2 \sin(\alpha L) = 0.$$

For a non-trivial solution ($C_2 \neq 0$), we require:

$$\sin(\alpha L) = 0 \quad \Rightarrow \quad \alpha L = n\pi, \quad n = 1, 2, 3, \dots$$

Thus, the permissible values of α (and hence λ) are:

$$\alpha_n = \frac{n\pi}{L} \quad \Rightarrow \quad \lambda_n = \alpha_n^2 = \left(\frac{n\pi}{L}\right)^2, \quad n = 1, 2, 3, \dots$$

The corresponding eigenfunctions are:

$$y_n(x) = C_2 \sin\left(\frac{n\pi}{L}x\right).$$

We can normalize the eigenfunctions by choosing $C_2 = 1$:

$$y_n(x) = \sin\left(\frac{n\pi}{L}x\right).$$

Case 2: $\lambda = 0$

The general solution is:
$$y(x) = C_1 x + C_2.$$

Applying the boundary conditions:

Boundary Condition at $x = 0$:

$$y(0) = C_2 = 0.$$

Thus, $C_2 = 0$, and the solution reduces to:

$$y(x) = C_1 x.$$

Boundary Condition at $x = L$:

$$y(L) = C_1 L = 0.$$

This implies $C_1 = 0$, leading to the trivial solution:

$$y(x) = 0.$$

Therefore, $\lambda = 0$ does not yield a non-trivial solution.

Case 3: $\lambda = -\beta^2 < 0$

The general solution is:

$$y(x) = C_1 e^{\beta x} + C_2 e^{-\beta x}.$$

Applying the boundary conditions:

Boundary Condition at $x = 0$:

$$y(0) = C_1 + C_2 = 0 \quad \Rightarrow \quad C_1 = -C_2.$$

Thus, the solution becomes:

$$y(x) = C_2 \left(e^{-\beta x} - e^{\beta x} \right) = C_2 \left(-2 \sinh(\beta x) \right),$$

where we have used the identity $e^{\beta x} - e^{-\beta x} = 2 \sinh(\beta x)$.

Boundary Condition at $x = L$:

$$y(L) = -2 C_2 \sinh(\beta L) = 0.$$

For a non-trivial solution ($C_2 \neq 0$), we require:

$$\sinh(\beta L) = 0.$$

However, $\sinh(\beta L) > 0$ for all $\beta > 0$, so there are no non-trivial solutions in this case.

Only positive values of λ yield non-trivial solutions that satisfy the boundary conditions. Specifically, the permissible eigenvalues and corresponding eigenfunctions are:

$$\lambda_n = \left(\frac{n\pi}{L} \right)^2, \quad n = 1, 2, 3, \ldots$$

$$y_n(x) = \sin\left(\frac{n\pi}{L}x\right), \quad n = 1, 2, 3, \ldots$$

These eigenvalues and eigenfunctions form an infinite discrete set, which are fundamental in various applications such as vibration analysis, heat conduction, and quantum mechanics.

Verification

To verify that these solutions satisfy both the differential equation and the boundary conditions, consider a specific eigenvalue $\lambda_n = \left(\frac{n\pi}{L}\right)^2$ and its corresponding eigenfunction $y_n(x) = \sin\left(\frac{n\pi}{L}x\right)$.

Differential Equation:

$$y_n''(x) + \lambda_n y_n(x) = \frac{d^2}{dx^2}\sin\left(\frac{n\pi}{L}x\right) + \left(\frac{n\pi}{L}\right)^2 \sin\left(\frac{n\pi}{L}x\right).$$

Calculating the second derivative:

$$y_n''(x) = -\left(\frac{n\pi}{L}\right)^2 \sin\left(\frac{n\pi}{L}x\right).$$

Substituting back:

$$y_n''(x) + \lambda_n y_n(x) = -\left(\frac{n\pi}{L}\right)^2 \sin\left(\frac{n\pi}{L}x\right) + \left(\frac{n\pi}{L}\right)^2 \sin\left(\frac{n\pi}{L}x\right) = 0.$$

Thus, the differential equation is satisfied.

Boundary Conditions:

$$y_n(0) = \sin(0) = 0,$$

$$y_n(L) = \sin(n\pi) = 0.$$

Both boundary conditions are satisfied for all integers n.

Orthogonality of Eigenfunctions

An important property of the eigenfunctions $y_n(x) = \sin\left(\frac{n\pi}{L}x\right)$ is their orthogonality over the interval $[0, L]$. Specifically,

$$\int_0^L y_n(x)y_m(x)\,dx = 0 \quad \text{for} \quad n \neq m.$$

This orthogonality property is fundamental in expanding arbitrary functions in terms of the eigenfunctions, a technique widely used in solving partial differential equations via separation of variables.

Conclusion and Summary

The boundary value problem

$$y'' + \lambda y = 0, \quad y(0) = 0, \quad y(L) = 0$$

has non-trivial solutions only for positive eigenvalues $\lambda_n = \left(\frac{n\pi}{L}\right)^2$, where n is a positive integer. The corresponding eigenfunctions are sinusoidal functions of the form

$$y_n(x) = \sin\left(\frac{n\pi}{L}x\right).$$

These solutions are essential in various applications across physics and engineering, where they describe standing wave patterns, heat distribution, and quantum states, among others.

Final Solution

Therefore, the complete set of solutions to the boundary value problem is given by

$$\lambda_n = \left(\frac{n\pi}{L}\right)^2, \quad n = 1, 2, 3, \ldots$$

$$y_n(x) = \sin\left(\frac{n\pi}{L}x\right), \quad n = 1, 2, 3, \ldots$$

These eigenvalues and eigenfunctions satisfy both the differential equation and the boundary conditions, providing a comprehensive solution to the problem.

Graphical Representation

For visualization, consider plotting the first few eigenfunctions for $n = 1, 2, 3$:

The plot illustrates the sinusoidal nature of the eigenfunctions, with each successive eigenfunction having an additional half-wavelength fitting into the interval $[0, L]$.

First Three Eigenfunctions

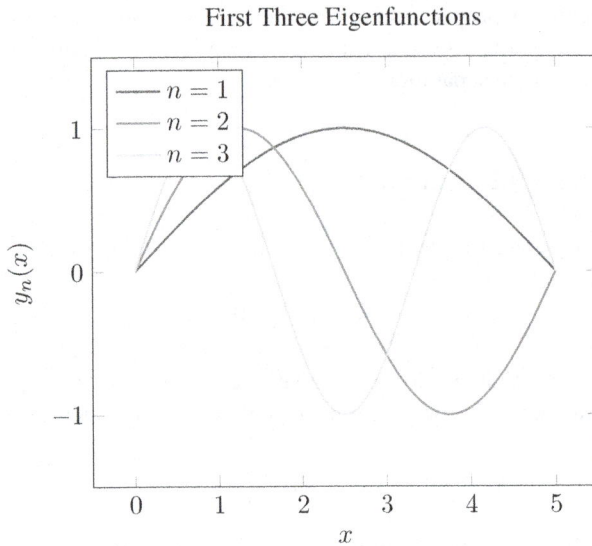

Figure 10.1: Eigenfunctions $y_n(x) = \sin\left(\frac{n\pi}{L}x\right)$ for $n = 1, 2, 3$ with $L = 5$.

Applications

The solutions obtained for this boundary value problem are fundamental in various physical contexts:

- **Vibration Analysis:** Describes the modes of vibration of a string fixed at both ends.

- **Heat Conduction:** Represents steady-state temperature distributions in a rod with fixed temperatures at both ends.

- **Quantum Mechanics:** Corresponds to the stationary states of a particle in a one-dimensional infinite potential well.

- **Electrical Engineering:** Models standing wave patterns in transmission lines.

Understanding these solutions provides valuable insights into the behavior of physical systems under various constraints and boundary conditions.

Further Considerations

Normalization of Eigenfunctions

In many applications, particularly in quantum mechanics, it is important to normalize the eigenfunctions. The normalization condition is:

$$\int_0^L y_n(x) y_m(x)\, dx = \delta_{mn},$$

where δ_{mn} is the Kronecker delta, which is 1 if $m = n$ and 0 otherwise. For our eigenfunctions $y_n(x) = \sin\left(\frac{n\pi}{L}x\right)$, we compute:

$$\int_0^L \sin\left(\frac{n\pi}{L}x\right) \sin\left(\frac{m\pi}{L}x\right)\, dx = \begin{cases} \frac{L}{2} & \text{if } n = m, \\ 0 & \text{if } n \neq m. \end{cases}$$

Therefore, the normalized eigenfunctions are:

$$\tilde{y}_n(x) = \sqrt{\frac{2}{L}} \sin\left(\frac{n\pi}{L}x\right).$$

Expansion of Arbitrary Functions

Any sufficiently smooth function $f(x)$ defined on $[0, L]$ satisfying the boundary conditions can be expressed as a series expansion in terms of the eigenfunctions:

$$f(x) = \sum_{n=1}^{\infty} a_n \sin\left(\frac{n\pi}{L}x\right),$$

where the coefficients a_n are given by:

$$a_n = \frac{2}{L} \int_0^L f(x) \sin\left(\frac{n\pi}{L}x\right)\, dx.$$

This property is instrumental in solving partial differential equations using the method of separation of variables.

Final Remarks

The boundary value problem $y'' + \lambda y = 0$ with $y(0) = 0$ and $y(L) = 0$ serves as a quintessential example in the study of differential equations, illustrating the concepts of eigenvalues and eigenfunctions. The discrete spectrum of eigenvalues and the orthogonality of the corresponding eigenfunctions lay the groundwork for more advanced topics in mathematical physics and engineering disciplines.

10.10   Model the Population Growth with a Logistic Differential Equation

Population dynamics is a fundamental aspect of biological studies, where understanding how populations grow and interact with their environment is crucial. While the simplest model of population growth is the exponential growth model, it assumes unlimited resources and an unbounded environment, which is rarely realistic. To address these limitations, the **logistic differential equation** is employed to model population growth considering environmental carrying capacity.

Formulation of the Logistic Differential Equation

Let $P(t)$ denote the population at time t. The logistic model incorporates the concept of carrying capacity K, which is the maximum population size that the environment can sustain indefinitely. The intrinsic growth rate of the population is represented by r.

The logistic differential equation is given by:

$$\frac{dP}{dt} = rP\left(1 - \frac{P}{K}\right)$$

Explanation of Terms:

- $\frac{dP}{dt}$: The rate of change of the population with respect to time.

- rP: Represents the exponential growth component, where the population grows proportionally to its current size.

- $\left(1 - \frac{P}{K}\right)$: This term introduces the effect of limited resources, decreasing the growth rate as the population P approaches the carrying capacity K.

Solving the Logistic Differential Equation

To solve the logistic equation, we use the method of **separation of variables**, which involves rearranging the equation to separate the variables P and t on opposite sides of the equation.

Step 1: Separation of Variables

Starting with the logistic equation:

$$\frac{dP}{dt} = rP\left(1 - \frac{P}{K}\right)$$

Rearrange to separate P and t:

$$\frac{dP}{P\left(1 - \frac{P}{K}\right)} = r\,dt$$

Step 2: Partial Fraction Decomposition

To integrate the left-hand side, we perform partial fraction decomposition. Let's express the integrand as a sum of simpler fractions:

$$\frac{1}{P\left(1 - \frac{P}{K}\right)} = \frac{A}{P} + \frac{B}{1 - \frac{P}{K}}$$

Multiplying both sides by $P\left(1 - \frac{P}{K}\right)$, we get:

$$1 = A\left(1 - \frac{P}{K}\right) + BP$$

Expanding and rearranging:

$$1 = A - \frac{A}{K}P + BP$$

Equating coefficients for like terms:

$$\text{Constant term:} \quad A = 1$$
$$\text{Coefficient of } P : \quad -\frac{A}{K} + B = 0$$

Substituting $A = 1$ into the second equation:

$$-\frac{1}{K} + B = 0$$
$$\Rightarrow B = \frac{1}{K}$$

Thus, the partial fractions are:

$$\frac{1}{P\left(1 - \frac{P}{K}\right)} = \frac{1}{P} + \frac{1/K}{1 - \frac{P}{K}}$$

Step 3: Integration

Substituting the partial fractions back into the separated equation:

$$\left(\frac{1}{P} + \frac{1/K}{1 - \frac{P}{K}}\right) dP = r\, dt$$

Integrate both sides:

$$\int \left(\frac{1}{P} + \frac{1/K}{1 - \frac{P}{K}}\right) dP = \int r\, dt$$

Calculating the integrals:

$$\int \frac{1}{P}\, dP = \ln|P| + C_1$$

$$\int \frac{1/K}{1 - \frac{P}{K}}\, dP = -\ln\left|1 - \frac{P}{K}\right| + C_2$$

$$\int r\, dt = rt + C_3$$

Combining the constants of integration ($C_1 - C_2 = C$):

$$\ln|P| - \ln\left|1 - \frac{P}{K}\right| = rt + C$$

Step 4: Solving for $P(t)$

Exponentiate both sides to eliminate the natural logarithm:

$$e^{\ln\left(\frac{P}{1 - \frac{P}{K}}\right)} = e^{rt+C}$$

Simplifying:

$$\frac{P}{1 - \frac{P}{K}} = Ce^{rt}$$

where $C = e^C$ is a constant.

228

Rearrange to solve for P:

$$P = Ce^{rt}\left(1 - \frac{P}{K}\right)$$

$$P = Ce^{rt} - \frac{C}{K}e^{rt}P$$

$$P + \frac{C}{K}e^{rt}P = Ce^{rt}$$

$$P\left(1 + \frac{C}{K}e^{rt}\right) = Ce^{rt}$$

$$P = \frac{Ce^{rt}}{1 + \frac{C}{K}e^{rt}}$$

$$P = \frac{KCe^{rt}}{K + Ce^{rt}}$$

To determine the constant C, we use the initial condition $P(0) = P_0$:

$$P(0) = \frac{KCe^{r \cdot 0}}{K + Ce^{r \cdot 0}}$$

$$P_0 = \frac{KC}{K + C}$$

$$P_0(K + C) = KC$$

$$P_0K + P_0C = KC$$

$$P_0K = KC - P_0C$$

$$P_0K = C(K - P_0)$$

$$C = \frac{P_0K}{K - P_0}$$

Substituting C back into the expression for $P(t)$:

$$P(t) = \frac{K\left(\frac{P_0K}{K - P_0}\right)e^{rt}}{K + \left(\frac{P_0K}{K - P_0}\right)e^{rt}}$$

$$P(t) = \frac{KP_0e^{rt}}{(K - P_0) + P_0e^{rt}}$$

$$P(t) = \frac{K}{1 + \left(\frac{K}{P_0} - 1\right)e^{-rt}}$$

Final Solution

229

The population at time t is given by the logistic function:

$$P(t) = \frac{K}{1 + \left(\frac{K}{P_0} - 1\right) e^{-rt}}$$

Analysis of the Solution

The logistic model exhibits several key behaviors:

- **Initial Growth:** When P is much smaller than K, the term $\frac{P}{K}$ is negligible, and the equation approximates exponential growth:

$$P(t) \approx P_0 e^{rt}$$

- **Carrying Capacity:** As t approaches infinity, e^{-rt} approaches zero, and the population stabilizes at the carrying capacity:

$$\lim_{t \to \infty} P(t) = K$$

- **Inflection Point:** The population growth rate is maximized when $P = \frac{K}{2}$, where the curve changes concavity.

Graphical Representation

TheFigure below illustrates the characteristic S-shaped logistic growth curve, where the population grows rapidly initially and then slows as it approaches the carrying capacity K.*

The logistic differential equation provides a more realistic model for population growth by incorporating the effects of limited resources through the carrying capacity K. By solving the logistic equation, we obtain a solution that describes how a population grows rapidly when it is small and then stabilizes as it reaches the environmental limits. This model is widely applicable in ecology, biology, and various fields where growth processes are subject to constraints.

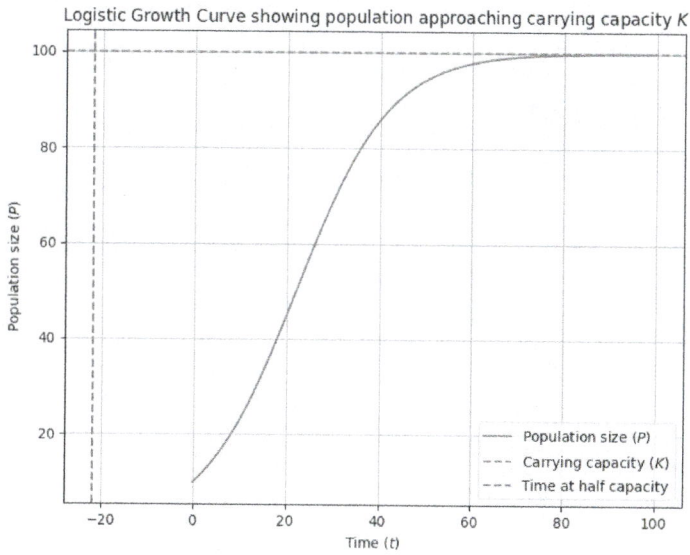

Logistic Growth Curve showing population approaching carrying capacity K

Made in the USA
Monee, IL
08 January 2025

76388564R00128